Acute Blood Purification

Contributions to Nephrology

Vol. 166

Series Editor

Claudio Ronco Vicenza

Acute Blood Purification

Volume Editors

Hiromichi Suzuki Saitama
Hiroyuki Hirasawa Chiba

38 figures, 5 in color, and 31 tables, 2010

Basel · Freiburg · Paris · London · New York · Bangalore · Bangkok · Shanghai · Singapore · Tokyo · Sydney

Contributions to Nephrology
(Founded 1975 by Geoffrey M. Berlyne)

Hiromichi Suzuki
Department of Nephrology
Saitama Medical University
Morohongo 38,
Moroyamamachi
Irumagun
Saitama 350-0495
Japan

Hiroyuki Hirasawa
Department of Emergency and
Critical Care Medicine
Chiba University
Graduate School of Medicine
1-8-1 Inohana, Chuo
Chiba 260-8677
Japan

Library of Congress Cataloging-in-Publication Data

Acute blood purification / volume editors, Hiromichi Suzuki, Hiroyuki Hirasawa.
p. ; cm. -- (Contributions to nephrology, ISSN 0302-5144; v.166)
Includes bibliographical references and indexes.
ISBN 978-3-8055-9478-3 (hard cover : alk. paper)
1. Blood--Filtration--Japan. 2. Critical care medicine--Japan. 3. Acute renal failure--Treatment--Japan. I. Suzuki, Hiromichi, M.D. II. Hirasawa, Hiroyuki, 1940- III. Series: Contributions to nephrology, v.166. 0302-5144;
[DNLM: 1. Renal Dialysis--methods--Japan. 2. Renal Replacement Therapy--methods--Japan. 3. Acute Disease--therapy--Japan. 4. Critical Care--methods--Japan. 5. Kidney Failure, Acute--therapy--Japan. W1 CO778UN v.166 2010 / WJ 378 A1885 2010]
RC901.7.H47A28 2010
362.17'840952--dc22

2010010692

Bibliographic Indices. This publication is listed in bibliographic services, including Current Contents® and Index Medicus.

www.karger.com
Printed in Switzerland on acid-free and non-aging paper (ISO 9706) by Reinhardt Druck, Basel
ISSN 0302–5144
ISBN 978–3–8055–9478–3
e-ISBN 978–3–8055–9479–0

Contents

Preface

Recently, it has become well known that the kidney plays the vital role of water-electrolyte regulation in critically ill patients. Various cytokines are involved in the pathophysiological process of acute organ damage and subsequently multiple organ failure. These overwhelming cytokines are not eliminated even with the kidneys functioning at their maximum capacity.

To prevent these catastrophic events, an innovative concept of acute blood purification has been developed in Japan. However, few physicians working in the field of critical care medicine in the world understand and apply this approach and technology to patients suffering from severe organ damage. This book describes the present status of acute blood purification in Japan. It is hoped that readers will come to understand the concept of elimination of cytokines in patients with septic shock in which cytokine storms produce serious organ damage, and the application of the technology of hemodiafiltration for elimination of cytokines. After considering the concept and technology, we will discuss how to construct the system of acute blood purification which includes various machines, devices, membranes, fluids and so on. In addition to these special aspects of acute blood purification, as the tool for the standard care of many critically ill patients with severe acute kidney injury, the role of continuous renal replacement therapy is discussed by leading experts in this field.

We thank all of the authors for their efforts to convey their knowledge in a concise but complete text, and to Karger Publishers for the efficient publication of the present volume.

H. Hirasawa, Chiba
H. Suzuki, Saitama

Suzuki H, Hirasawa H (eds): Acute Blood Purification.
Contrib Nephrol. Basel, Karger, 2010, vol 166, pp 1–3

Introduction

Claudio Ronco

Department of Nephrology, San Bortolo Hospital, International Renal Research Institute, Vicenza, Italy

For many years, the term 'blood purification' has been used to indicate renal replacement therapy directed at chronic patients with end-stage kidney disease. The level of application and understanding of extracorporeal therapies for renal replacement and support have however been expanded in recent years, and today a new area of clinical application and research is the use of blood purification techniques in the critically ill. This evolution has required an expansion of the multidisciplinary approach to critical care and nephrology, resulting in a brand new speciality called 'critical care nephrology' [1]. At the same time, the multiple applications of extracorporeal therapies in critically ill patients have made possible the evolution of management of such patients using a new therapeutic strategy called 'multiple organ support therapy' [1–3]. In these circumstances, the extracorporeal therapies are called upon to support many organs other than the kidney and for more than just the manipulation or correction of circulating blood composition.

For many years, renal replacement therapy has been a technique used by nephrologists, while intensive care and nephrology were regarded as two separate specialities. Different training as well as different clinical views were part of the organ-oriented education. This has somehow led to a separation of the competences and a separation of the clinical approaches, taking medical practice far from the holistic and patient-oriented approach. Today, the complexity of acute kidney injury syndromes, the concomitant presence of sepsis and the frequent occurrence of multiple organ dysfunction represent a condition in which severity of disease must be approached by a multidisciplinary task force with a high level of clinical and logistical integration (fig. 1). Patients may have high severity scores, be on different life support systems and receive multiple pharmacological support therapies; thus, acute kidney injury represents only part of the clinical picture and the modern physician must be prepared to tackle multiple and complex clinical problems. Furthermore, this new approach has required an evolution of the applied technology, moving from the traditional 'dialysis machines' to more specific equipment

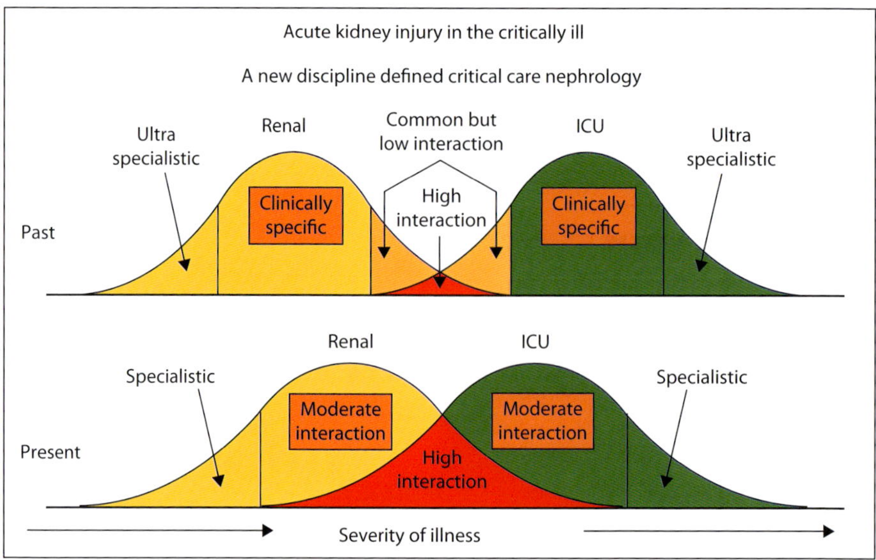

Fig. 1. In the past, the interaction between nephrology and intensive care was minimal. Today, there is continuous interaction with several moments of high interaction due to common patients and complex syndromes.

for acute patients. Critically ill patients require a level of monitoring and care that is more intensive and energy consuming compared to the chronic patient. While in chronic dialysis one nurse can take care of patients at a ratio of 1:4, in intensive care, especially in presence of mechanically ventilated patients, this ratio must go to 1:1. When treating patients at risk of multiple organ failure that requires renal replacement and support, physicians require a full understanding of the implications that an extracorporeal therapy may have on other organs. Furthermore, not all extracorporeal therapies are equal, and therefore careful use of correct terminology is strongly recommended [2]. Considering the difference between renal replacement and renal support, we may have absolute and relative indications for the application of extracorporeal therapies. Furthermore, when sepsis and septic shock are present, adjunctive technologies may be utilized, making the extracorporeal circuit a tool for correction of homeostatic imbalances, intoxication and immunodysregulation [3]. Intensivists and nephrologists must work together to provide the best possible care. Vascular access, anticoagulation strategies, type of membranes and solutions must be chosen in agreement with the general requirements of the critically ill patient. This is even more true when a pediatric population is involved. New equipment for continuous renal replacement therapies has enormous potential for different treatment modalities and for long-lasting treatments with minimal technical and clinical complications. Nevertheless, a strong emphasis should be placed on education and training of the personnel to ensure that a theoretically adequate prescription results in an effective delivery of the prescribed therapy. Timing of intervention, dose of treatment, and modality of solute and fluid removal represent an important input that physicians must provide at the moment of initiation [4]. Accurate monitoring, appropriate maneuvers and optimal care of the vascular access and the extracorporeal circuit represent important aspects of the nursing care. Both components are quintessential in making a therapy the most safe, efficient and well-tolerated for the fragile and complicated critically ill patient. We are facing a new era of medicine in which the knowledge of the individual must be integrated with the performance of the entire team. Only by approaching the complexity of the new syndromes through a new multi-disciplined yet efficient task force may we provide new hope to critically ill patients suffering from renal disorders who may have other concomitant problems and significant comorbidities. This will be a challenge for years to come.

References

1 Ronco C, Bellomo R: Critical care nephrology: the time has come. Nephrol Dial Transplant 1998;13:264–267.

2 Ronco C, Bellomo R: Continuous renal replacement therapy: evolution in technology and current nomenclature. Kidney Int 1998;53(suppl 66):S160–S164.

3 Ronco C, Bellomo R: Acute renal failure and multiple organ dysfunction in the ICU: from renal replacement therapy (RRT) to multiple organ support therapy (MOST). Int J Artif Organs 2002;25:733–747.

4 Bagshaw SM, Cruz DN, Gibney RT, Ronco C: A proposed algorithm for initiation of renal replacement therapy in adult critically ill patients. Crit Care 2009;13:317.

Claudio Ronco, MD
Department of Nephrology, San Bortolo Hospital, International Renal Research Institute
IT–36100 Vicenza (Italy)
Tel. +39 0444753650, Fax +39 0444753949, E-Mail cronco@goldnet.it

Suzuki H, Hirasawa H (eds): Acute Blood Purification.
Contrib Nephrol. Basel, Karger, 2010, vol 166, pp 4–10

Current Status of Blood Purification in Critical Care in Japan

Kazo Kaizu · Yoshifumi Inada · Akio Kawamura · Seito Oda · Hiroyuki Hirasawa

Survey Committee, Japan Society for Blood Purification in Critical Care, Yokohama, Japan

Abstract

In order to clarify the present status of blood purification therapy (BPT) in critical care in Japan, questionnaires investigating all the patients who were treated with BPT in 2005 were distributed. The number of patients who received BPT was 9,795, and the number of BPT performed was 11,623. The number and types of BPT treatment given are: continuous hemodiafiltration (CHDF)/hemofiltration (HDF) 5,443 (50.3%); continuous hemofiltration (CHF) 812 (7.5%); continuous hemodialysis (CHD) 877 (8.1%); simple plasma exchange 898 (8.3%); direct hemoperfusion (DHP) with polymyxin-B-coated textile (PMX-DHP) 1,625 (15.0%); DHP with activated carbon (AC-DHP) 129 (1.2%). The survival rates of patients with continuous therapies (CHDF, CHF, CHD) were as follows: multiple organ failure with CHDF 35%; sepsis with CHDF 65%; acute hepatic failure with CHDF 50%; acute renal failure with CHDF 66%; acute drug intoxication with AC-DHP 79%. In conclusion, continuous therapies such as CHDF, CHF and CHD were the most popular modes (>65%) of BPT in Japan. The worst survival rate among diseases in critical care was found in multiple organ failure patients. The best survival rate was in those who suffered from acute renal failure.

Continuous hemofiltration (CHF), or continuous renal replacement therapy, has undergone remarkable growth [1]. Recently, blood purification therapy (BPT) has not only developed markedly by using new technology [2–4], but has also been distributed widely and rapidly all over Japan. BPT has been used to treat many patients with varying diseases. An increase in the number of clinical engineers (specialists in operating the many kinds of sophisticated machines used in BPT) may have contributed to the development of this therapy.

Table 1. BPT currently used in Japan with common abbreviations

Continuous hemodiafiltration	CHDF
Hemofiltration	HF
Continuous hemofiltration	CHF
Continuous hemodialysis	CHD
Simple plasma exchange	SPE
Direct hemoperfusion with polymyxin-B coated textile	PMX-DHP
Direct hemoperfusion with activated carbon	AC-DHP
Immunoadsorption therapy	IA
Others (L-CAP, G-CAP)	

The aim of this study was to clarify the numbers of both patients treated with BPT and BPT treatments themselves. In addition, both efficacies and survival rates of patients using BPT were also examined.

Methods

Modes of BPT

There are many modes of BPT which were available in Japan in 2005 (table 1).

Diseases Requiring BPT

The diseases treated with BPT in Japan are as follows: (1) multiple organ failure, (2) acute renal failure, (3) acute exacerbation of chronic renal failure, (4) acute hepatic failure (including postoperative acute hepatic failure), (5) sepsis (including endotoxemia), (6) acute respiratory distress syndrome, (7) severe acute pancreatitis, (8) congestive heart failure, (9) acute drug intoxication, (10) acute disorders of electrolytes, water and acid-base of blood; (11) acute metabolic disorder/congenital metabolic disorder, (12) thrombotic thrombocytopenic purpura/hemolytic uremic syndrome (TTP/HUS), (13) acute exacerbation of autoimmunological disease, (14) Guillain-Barré syndrome, (15) thyroid crisis, and (16) toxic epidermal necrolysis.

Questionnaires

This survey was carried out by the analysis of questionnaires which were sent by mail to doctors who were members of the Japan Society for Blood Purification in Critical Care, and Japanese Society for Intensive Care Medicine, and chief doctors of dialysis departments of university hospitals in Japan.

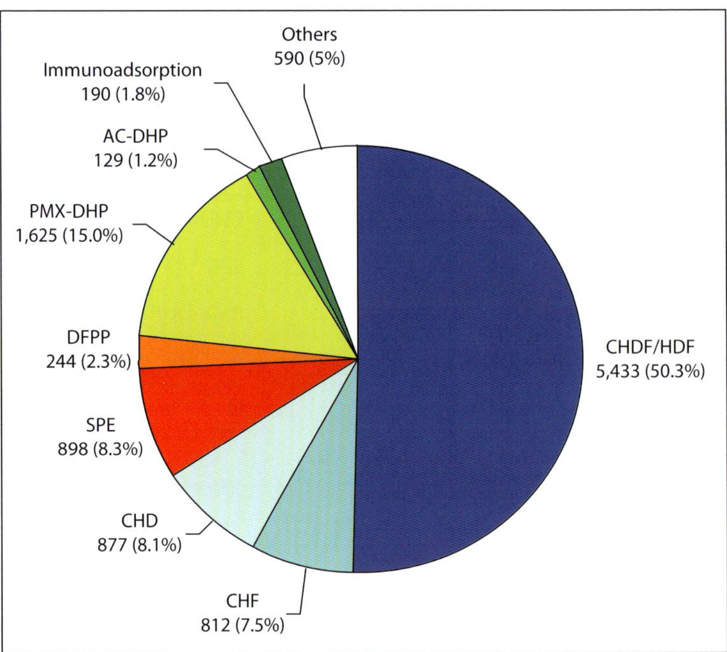

Fig. 1. Numbers of BPT performed.

Results

In 2005, 9,795 patients were treated with BPT (males: 6,295, mean age: 63.3 ± 17.9 years; females: 3,500, mean age: 62.4 ± 20.0 years; unknown gender: 7). In total, 11,623 BPT were performed.

Modes of BPT

The numbers of each BPT treatment are as follows: continuous hemodiafiltration/hemofiltration (CHDF/HDF) 5,433 (50.3%); CHF 812 (7.5%); continuous hemodialysis (CHD) 877 (8.1%); SPE 1,142 (10.6%); direct hemoperfusion with polymyxin-B-coated textile (PMX-DHP) 1,625 (15.0%); AC-DHP 129 (1.2%); immunoadsorption therapy 190 (1.8%) (fig. 1).

The numbers of each disease treated with BPT are as follows: multiple organ failure 1,409 (12.0%); acute renal failure 1,993 (17.0%), acute exacerbation of chronic renal failure 1,160 (10%); congestive heart failure 1,130 (10%); acute disorders of electrolyte, water and acid-base of blood (347 (3.0%); severe acute pancreatitis 236 (2.0%); acute hepatic failure 709 (6.0%); TTP/HUS 208 (2.0%); sepsis 2,096 (18.0%); acute respiratory distress syndrome 160 (1.0%); SLE-MG (Systemic Lupus Erythematosus-Myasthenia Gravis)

363 (3.0%); Guillain-Barré syndrome 169 (1.5%); acute drug intoxication 163 (1.4%); MD (Miscellaneous Disease) 5 (0.004%). In multiple organ failure, CHDF/HDF (73%) was the most popular mode used, followed by CHD (8%) and CHF (7%).

In acute renal failure, the most popular mode was CHDF (76%), followed by CHD (10%) and CHF (10%). In severe acute pancreatitis, the most popular mode was also CHDF (78.0%), followed by CHF (11.0%). In acute hepatic failure, the most popular mode was simple plasma exchange (SPE; 30%), followed by CHDF/HDF (23.0%). In TTP/HUS, the most popular mode was SPE (64%), followed by double filtration of plasmapheresis. In sepsis, PMX-DHP was the most popular mode (67%), followed by CHDF/HDF (24%). In acute respiratory distress syndrome, CHDF/HDF was the most popular mode (64%), followed by PMX-DHP (18%).

Efficacy and Survival (Fig. 2, 3)

The efficacy of CHDF and survival rates of patients with different numbers of failure organs, respectively, are as follows: in patients with single-organ failure 71.8 and 58.8%; 2-organ failure 45.5 and 36.4%; 3-organ failure 54.9 and 33.3%; 4-organ failure 27.8 and 22.2%; 5-organ failure 42.9 and 28.6%. There was a significant difference between single-organ failure and the others. Efficacy of BPT and survival rates of multiple-organ failure patients treated with CHDF, CHF, CHD and HDF are: 48 and 33% in CHDF; 40 and 40% in CHF; 33 and 17% in CHD; 60 and 30% in HDF. There was no difference among these groups. Efficacy of BPT and the survival rates of septic patients treated with PMX-DHP and/or CHDF/HDF were as follows: 72 and 70% in PMX-DHP alone, 64 and 60% in CHDF+PMX-DHP, and 56 and 36% in CHDF alone. There was a significant difference in survival rates between CHDF+PMX-DHP and CHDF alone, although no difference in efficacies among 3 groups was found. Efficacy of BPT and survival rates of patients with acute hepatic failure treated with SPE and/or CHDF were: 70.6 and 76.5% in SPE alone; 70.2 and 45.6% in CHDF+SPE; 54.2 and 37.5% in CHDF alone. Survival rates in SPE were significantly different from those in CHDF+SPE and CHDF alone. Efficacy of BPT and survival rates of patients with acute drug-intoxication treated with CHDF, AC-DHP and HDF were: 83 and 33% in CHDF; 89 and 88% in AC-DHP; 100 and 100% in HDF. There were no differences in efficacy and survival rates among the groups.

Discussion

This is the first nationwide study on BPT in critical care in Japan. The total number of patients who received BPT in 2005 was 9,846. Since some patients received more than one mode of BPT, the total number of BPT performed was 11,623. Since the response rate for the questionnaire was about 50%, the

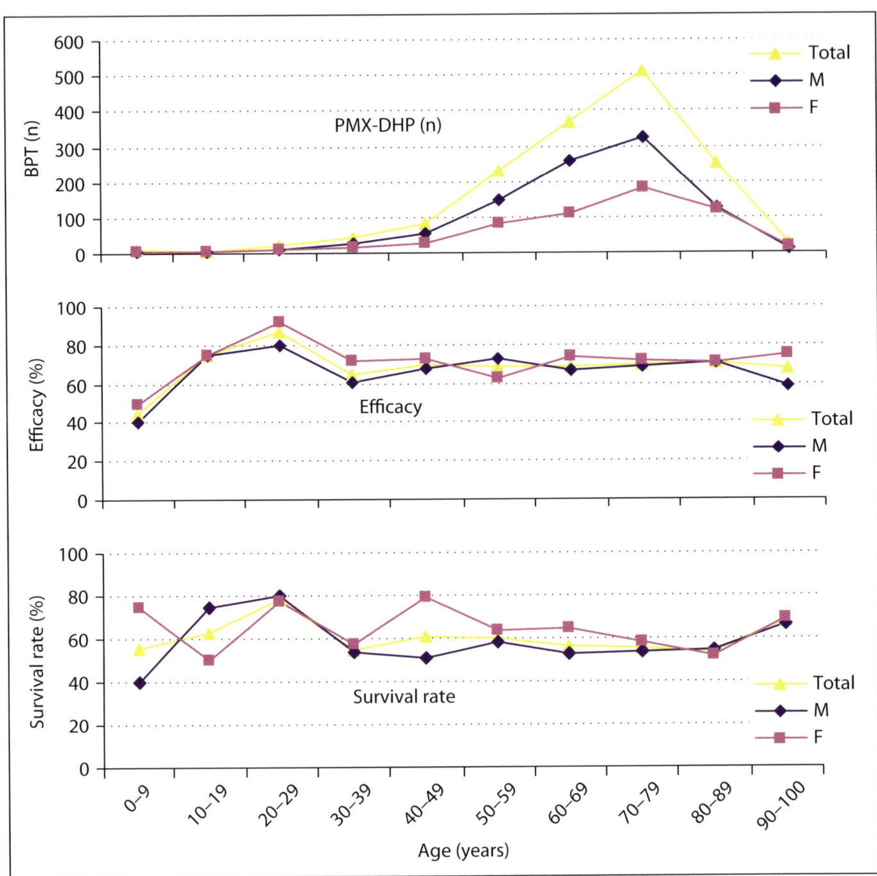

Fig. 2. Efficacy and survival in patients treated with CHDF/HDF.

real numbers of BPT performed and patients with BPT must be much greater than suggested by our data. Among all the modes of BPT, continuous therapies, such as CHDF, CHD and CHF, were the most used (61.9%). In particular, CHDF made up 50.3% (5,433) of all the BPT. PMX-DHP, which was first created in Japan, was performed quite often (15%, 1,625) in patients with sepsis, although it is very expensive. However, the survival rates which were calculated here are not accurate as this study was not a randomized controlled trial. Although this survey was a nationwide study, the recovery rate was only 50.7%; therefore, it cannot provide a complete picture of the current status of BPT in Japan. Randomization would have to be stratified according to illness severity, cause of acute renal failure and hospital [5]. So far, no accurate survival rate and efficacy of BPT including continuous renal replacement therapy has been presented [5].

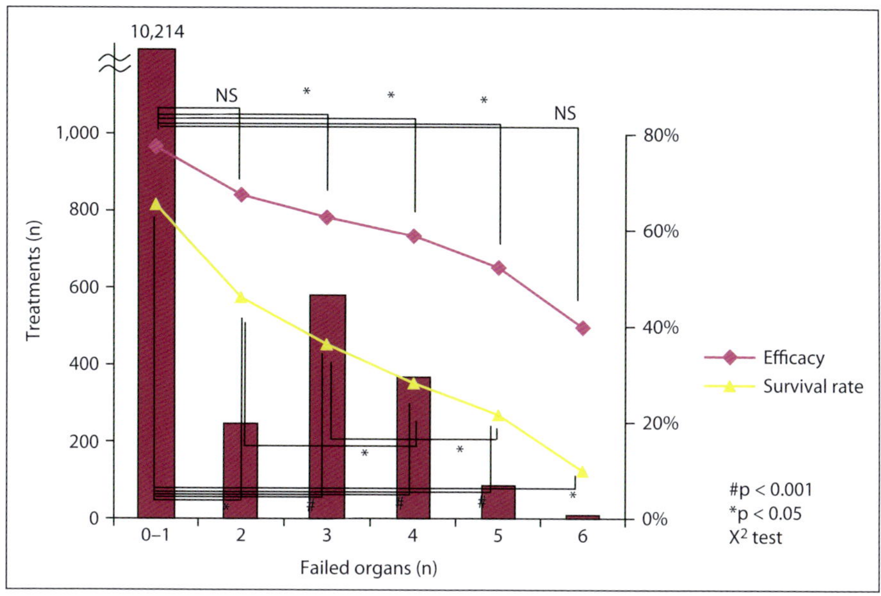

Fig. 3. Efficacy and survival in patients with different numbers of organ failure.

We have to understand that these results represent the minimum number of BPT performed in Japan. Another important point concerns the efficacy of treatment, since this was judged by the treating physician, it is less objective.

In conclusion, the number of BPT performed in Japan is both large and on the increase.

Acknowledgement

We sincerely thank Mrs. Yasuko Tanaka for her assistance.

References

1 Burchardi H: History and development of continuous renal replacement techniques. Kid Int Suppl 1998;53(suppl 66):s12–s124.

2 Bellomo R, Ronco C: Continuous versus intermittent renal replacement therapy in the intensive care unit. Kid Int Suppl 1998;53(suppl 66)s125–s128.

3 Bellomo R, Tipping P, Boyce N: Continuous veno venous hemofiltration with dialysis removes cytokines from the circulation of septic patients. Crit Care Med 1993;21:522–526.

4 Grootendost AF, van Bommel FH, van del Hoven B, van Leengoed LA, van Osia AL: High volume hemofiltration improves right ventricular function in endotoxin-induced shock in the pig. Intensive Care Med 1992;18:235–240.

5 Bellomo R, Ronco C: Continuous hemofiltration in the intensive care unit. Critical Care 2000;14:339–345.

Kazo Kaizu
Department of Nephrology & Blood Purification
Social Insurance Yokohama Chuo Hospital
268 Yamashita-cho Naka-ku, Yokohama 231-8553, Japan
Tel +81 045 641 1921, Fax +81 045 671 9872, E-Mail kaizu.kazo@yokochu.jp

Suzuki H, Hirasawa H (eds): Acute Blood Purification.
Contrib Nephrol. Basel, Karger, 2010, vol 166, pp 11–20

Terminology and Classification of Blood Purification in Critical Care in Japan

Hideki Kawanishi

Tsuchiya General Hospital, Hiroshima, Japan

Abstract

Blood purification in critical care (BPCC) has many indications. Acute kidney injury is a major indication, but there are also non-renal indications that are frequently complicated by kidney injuries. BPCC is performed not only by hemodialysis or hemofiltration using hemofilters, but also various other methods such as apheresis and adsorption. Indications for such a wide range of therapeutic options must be evaluated. Recently, the standardization of terminology and definitions has been attempted (primarily by American and European nephrologists and intensive care physicians) and a standardization committee was organized. The results of this evaluation were made public by the Acute Dialysis Quality Initiative. The Japan Society for Blood Purification in Critical Care has also proposed a standardization of the terminology. Such terminology and definitions should be utilized when making presentations or publishing papers.

Due to developments in both the equipment used and skills required, blood purification has now been classified into apheresis, adsorption and hemofiltration (HF)/hemodialysis (HD). Its application has widened from supporting kidney function to treatment for liver failure and sepsis (fig. 1). However, the complicated nature of these diseases requires the simultaneous use of multiple techniques, which often causes confusion with their classification.

Some terms related to blood purification are derived from chronic dialysis therapy, but others are from intensive care. There are also various abbreviations that have not yet been standardized. In addition, nephrologists and intensive care physicians who are primarily involved in blood purification have different educational backgrounds, and communication between them has often been inadequate. Recently, however, the Acute Dialysis Quality Initiative (www.adqi.net) has been established as an international organization to standardize the definitions and terminology related to therapy. In Japan, the Japan Society

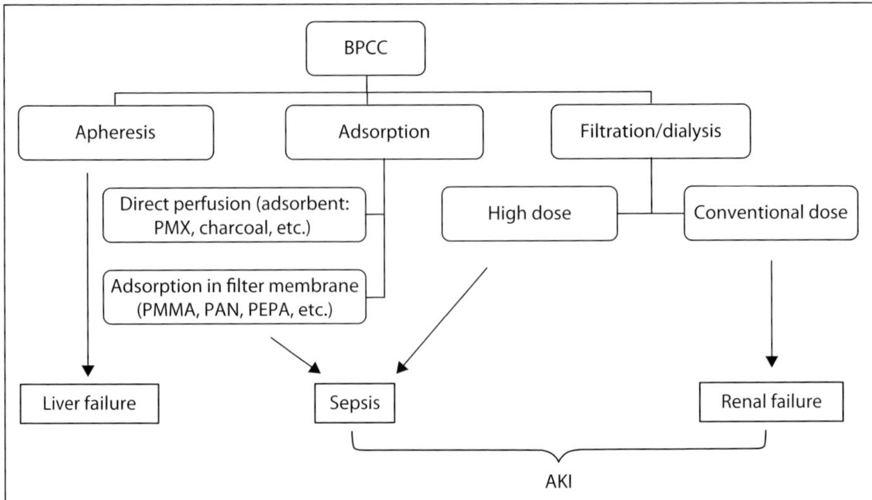

Fig. 1. Classification of blood purification in critical care (BPCC) technology. PMX = polymyxin-B immobilized fiber; PMMA = polymethylmethacrylate; PAN = polyacrylonitrile; PEPA = polyether polymer alloy.

for Blood Purification in Critical Care (JSBPCC) also drafted the 'Terminology concerning blood purification in critical care' in 2003 (http://jsbpcc.umin.jp) [1]. In our article, BPCC terminology and its classification in Japan have been explained using the JSBPCC glossary. Since this article focuses on terms related to therapies using HF/HD, those related to apheresis or adsorption have been excluded.

Expressions of Blood Purification in Critical Care

Expressions describing blood purification used to treat acute diseases have not been standardized; however, the JSBPCC refers to 'blood purification in critical care' (BPCC), i.e. it assumes that it is used in intensive care.

Many diseases that require BPCC are complicated by acute kidney injury (AKI). Therefore, they are all usually categorized as AKI in western countries, where the expression 'renal replacement therapy' (RRT) is prevalent.

In Japan, the term RRT is also often used in the field of chronic dialysis. However, it is not common in the field of intensive care. Also, as liver failure and acute pancreatitis not complicated by kidney failure are also treated by employing this technique, the JSBPCC uses 'blood purification' as a comprehensive term. Recently, blood purification has often been performed to control inflammatory mediators (non-renal indications) as well as for kidney support [2, 3].

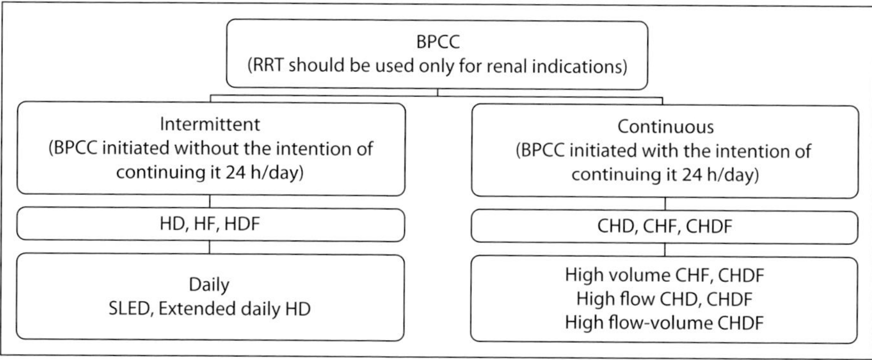

Fig. 2. The calcification of terminology of blood purification in critical care. HDF = hemodiafiltration.

Therefore, if the term RRT is used in the field of BPCC, it should be understood that non-renal indications are excluded.

Definitions of Continuous and Intermittent Therapies

Intermittent blood purification was used to mean blood purification every other day, similar to normal chronic HD. However, with the diversification of BPCC, daily and long-duration intermittent blood purification is now being chosen and the difference compared with continuous therapy has therefore diminished [4]. The following definitions have been established to clarify the differences (fig. 2):

Continuous Blood Purification

Blood purification initiated with the intention of continuing it for 24 h/day is defined as continuous blood purification, even if it has not been sustained for 24 h due to unavoidable circumstances. When it is performed, the methods and circumstances of its implementation [(hemofilter, blood flow (Q_B), dialysis fluid flow (Q_D), substitution fluid flow and filtration rate (Q_F)] must be recorded.

Intermittent Blood Purification

Blood purification that is not continuous is invariably classified as intermittent. Blood purification is intermittent if its interruption is anticipated at the onset. In administering intermittent blood purification, the method, duration and conditions of its implementation (hemofilter, Q_B, Q_D, substitution fluid flow, and Q_F) must be recorded.

It should be noted that the conditions of intermittent blood purification are important because it may be daily or sustained low-efficiency dialysis (SLED [5, 6]; also expressed as 'extended daily dialysis' [7, 8]). Incidentally, SLED is blood purification over 8–10 hours administered daily at a lower Q_D than in usual HD, and it is generally assumed to be performed during the day shift.

Classification Based on the Purification Mode Using a Hemofilter

Blood purification using hemofilters includes HD, HF and hemodiafiltration (HDF), but these 3 modes differ only in Q_D and Q_F, and should be regarded as continuous. Therefore, in BPCC, optimal therapeutic conditions must be maintained by constantly changing the Q_D and Q_F according to the patient's condition.

Historically, HD appeared first, and HF was subsequently developed as a convective type of artificial kidney due to improvements in dialyzers [9]. HF is a method of dialysis therapy in which solutes are removed from the blood by the convective transport of the filtrate, which is accomplished by applying a transmembrane pressure to the filter membrane, and the same volume of substitution fluid is supplemented. This therapy is based on a concept modeled after the glomerular function of the living kidney, and is effective for eliminating substances up to large solutes. Also, it is unlikely to cause disequilibrium syndrome, because there is no rapid change in the osmotic pressure. However, it does not replace the tubular function, and a substitution fluid containing electrolytes and an alkalizing agent must be administered to adjust the electrolyte levels and acid-base balance. HDF is a combination of HD and HF to compensate for the defects of HF, i.e. the inability to eliminate small solutes and adjust the electrolyte levels and acid-base balance.

In the field of chronic dialysis, HDF is regarded as a therapeutic mode devised to eliminate middle to large solutes by filtration through a combination of HD and HF, but it was initially developed by adding HD to HF for the elimination of small solutes, which cannot be accomplished sufficiently by HF [10]. In acute blood purification, filtration plays the primary role, and the addition of HD to HF resulted in the same therapy as the original hypothesis.

When these therapies are performed continuously, they are expressed as CHD, CHF and CHDF.

Classification of Blood Purification Based on the Blood Drawing and Returning Routes

Continuous blood purification began as continuous arteriovenous HF, in which water is removed continuously and slowly by the arteriovenous pressure gradient alone without a blood pump, and was reported in 1977 by Kramer et al. [11] in patients with heart failure/overhydration. Therefore, in western countries, many expressions based on the classification of continuous blood purification modes according to the blood drawing and returning routes have been used. Arteriovenous routes include: (1) continuous arteriovenous HF; (2) continuous arteriovenous HD; (3) continuous arteriovenous HDF. Venovenous routes include: (1) continuous venovenous HF; (2) continuous venovenous HD; (3) continuous venovenous HDF.

However, as blood is currently often transported using a blood pump (even in arteriovenous methods), attention to the arteriovenous pressure gradient has become unnecessary and the significance of indicating the blood drawing and returning routes has been lost. Thus, it has been recommended to avoid using complicated terms indicating these routes. However, indication of the blood drawing and returning routes is necessary when an arteriovenous method is deliberately performed without a blood pump.

Selection of CHD, CHF and CHDF Based on the Elimination Efficiency

Since the dialysis fluid/substitution fluid is supplied primarily by the bag fluid, the elimination efficiency is highest using CHF if the same volumes of dialysis fluid/substitution fluid are used by continuous blood purification, in which there is a dialysis dose limitation. However, in post-dilution CHF, the Q_F is dependent on the Q_B, and its limit is usually a quarter of the Q_B. The Q_F can be increased by pre-dilution CHF, but the increase in eliminated particles, particularly small solutes, decreases with increases in the Q_F (it is difficult to increase the Q_F if the replaced fluid volume is limited). Therefore, if post-dilution CHF is selected to eliminate substances up to large solutes, and CHD is added to increase the elimination of small solutes, this is termed CHDF.

In commonly practiced continuous blood purification, bag type dialysis fluid/substitution fluid is used, as mentioned previously. In this method, the Q_D and Q_F are ≤20 ml/min (being lower than the Q_B) and this flow rate determines the efficiency of the whole system. Factors that determine the efficiency in various modes are presented here.

Continuous Hemodialysis

Similarly to conventional HD, the clearance is determined by the least of the Q_D, Q_B and elimination performance of the hemofilter (KoA). In the continuous mode, since the Q_D is low ($Q_D < Q_B <$ KoA), the elimination of small solutes is determined by Q_D. In the elimination of large solutes ($Q_D <$ KoA $< Q_B$) it is also determined by the Q_D.

Continuous Hemofiltration
Similarly to conventional HF, the clearance is determined by Q_F.

Continuous Hemodiafiltration
If the Q_B is low, and $Q_D + Q_F$ is also low (<30 ml/min), Q_D and Q_F equally affect the clearance of small solutes. Therefore, the clearance of small solutes is equal to that of the dialysis fluid/substitution fluid in CHD, CHF and CHDF. The clearance of large solutes is determined by the Q_F, similarly to CHF.

However, using any method, time is an important factor in continuous treatment, and the amounts of eliminated solutes are calculated by multiplying these clearance values by time (min).

Expressions Used when the Volume of Blood Purification Is Increased Intentionally above the Standard Level

Terms concerning continuous therapy in which the volume of blood purification is increased are unclear. The Acute Dialysis Quality Initiative only shows 'continuous high-volume HF' and 'high-volume HF'. Terms that differentiate an increase in Q_F and Q_D are also unclear.

Therefore, the JSBPCC glossary defines therapy with an increased Q_F as 'high volume' and that with increased Q_D as 'high flow'. However, there is no established criterion that warrants the use of these terms. Presently, a Q_F of 35 ml/min/kg (≥2 l/h), at which a significant difference was demonstrated in the prospective study of Ronco et al. [12], is considered to be 'high', but evaluation of the results of future clinical studies is necessary.

High-Volume CHF
Defined as: CHF using a higher Q_F than normal. Although ≥2 l/h is considered appropriate as a criterion, a clear statement regarding the Q_F has been intentionally avoided. Note that the term 'high volume' is used when the filtration volume is increased (this also applies to the following terms).

High-Flow CHD
Defined as: CHD using a higher Q_D than normal; however, mention of a specific value has been intentionally avoided. Note that if the volume of dialysis fluid is increased, the term 'high flow' is used (this also applies to the following terms).

High-Volume CHDF
Defined as: CHDF using a higher filtrate flow rate than normal.

High-Flow CHDF
Defined as: CHDF using a higher Q_D than normal.

High-Flow/Volume CHDF
Defined as: CHDF using higher filtrate and Q_D than normal. Note that 'flow volume' is regarded as a single term.

Continuous High-Flux Dialysis
In this therapy, internal filtration-enhanced dialysis is applied to continuous dialysis therapy, and a highly permeable dialyzer is used with a blood and dialysis fluid countercurrent. Ultrafiltrate production is controlled by blood pumps, and there is a balance of filtration and backfiltration with ultrafiltrate produced in the proximal portion of the fibers and reinfused by backfiltration in the distal portion of the fibers, so that replacement fluid is not required [13, 14]. However, in continuous dialysis therapy, there is only a small amount of internal filtration, because the Q_B and Q_D are low. Therefore, it is usually used in intermittent therapy at the same Q_B and Q_D as those in regular dialysis therapy or SLED/extended daily HD therapy.

Expressions Based on the Administration Route of Substitution Fluid

In HF and HDF, the administration route of substitution fluid may be a blood-drawing route, blood-returning route, or both, and the elimination efficiency varies according to the route used. Therefore, the administration route must be indicated as follows: (1) pre-dilution, i.e. substitution fluid is administered from the inflow side of the blood purifier; (2) post-dilution, i.e. substitution fluid is administered from the outflow side of the blood purifier; (3) pre- and post-dilution (mixed-dilution), i.e. substitution fluid is administered from both the inflow and outflow sides of the blood purifier.

Terms used to Express Flow Rates

Abbreviations to express flow rates are given in capital letters, but lowercase letters are also used for convenience:

- Blood flow rate (Q_B, Qb).
- Dialysis fluid flow rate (Q_D, Qd).
- Substitution (replacement) flow rate (Q_S, Qs); note that in the field of chronic dialysis, the substitution flow rate is often expressed as Q_F, but preferably denoted as Q_S.
- Filtration flow rate (Q_F, Qf); note that the Q_F is the sum of the substitution flow and body fluid removal rates. In reality, this is the ultrafiltration rate

(UFR, QUF), but as the body fluid removal rate is often expressed as the ultrafiltration rate, the JSBPCC recommends not to use the ultrafiltration rate to avoid confusion.

- Body fluid removal rate (calculated by Q_F–Q_S)

Classification and Expressions of Hemofilters

In BPCC, the term 'hemofilter' is primarily used to refer to a blood purification device. In Japan, the price of 'continuous slow HF therapy' (JPY 19,900) was determined while reforming the medical insurance system in 2008. This is a term unique to the Japanese medical insurance system, and in fact includes CHD, CHF and CHDF. However, a continuous slow hemofilter approved as a special hemofilter for continuous slow HF therapy must be used to request payment. This restricts the variety of hemofilters used (if a dialyzer is used as a blood purification device, the the cost of chronic hemodialysis, JPY 15,900, is bound to be selected).

Presently, however, the difference in performance between hemofilters and dialyzers using the same membrane material is small, and the classification is simply based on the medical insurance system.

Therefore, when using a device approved for chronic hemodialysis as a hemofilter, it must be expressed as a dialyzer. In addition, the membrane type and area (or the product name and manufacturer) must also be indicated as essential information. Similarly, if a device approved as a hemofilter is used, it must be indicated as a hemofilter, and the membrane type and area (or the product name and manufacturer) are essential information.

Polysulfone, polyethersulfone, polyacrylonitrile, polyether polymer alloy, cellulose triacetate, and polymethylmethacrylate (PMMA) membranes are typical hemofilters. Polysulfone, polyethersulfone and cellulose triacetate membranes have low adsorption capacities, and remove substances primarily by diffusion and filtration. Therefore, no difference is observed in the elimination efficiency among the membrane types in BPCC with a restricted dialysis fluid/substitution fluid volume. However, the fiber diameter, length and fiber filling rate are modified by each manufacturer to improve the anti-thrombogenic properties.

PMMA membranes have a homogeneous membrane structure and are characterized by high protein adsorption [15]. Therefore, they can eliminate β_2-microblobulin and cytokines (such as IL-6) by adsorption. They are used for CHDF to control hypercytokinemia, such as that in sepsis [4, 16], but they are employed to remove cytokines by adsorption rather than filtration, so that the Q_F at which the adsorption capacity of the entire membrane can be maximized (<300 ml/h). One of their shortcomings is their poor anti-thrombogenic properties. In particular, the filter life is shortened at a high Q_F. Recently, the continuous use of a dialyzer with a large membrane area (>2.0 m^2) has been attempted as a method to increase the adsorption capacity of PMMA membranes.

Conclusion

In this section, basic terms related to BPCC were explained. These terms and their definitions are considered to be worth reviewing before making a presentation or writing a paper.

References

1 Japan Society for Blood Purification in Critical Care: Terminology concerning blood purification in critical care in 2003. http://jsbpcc.umin.jp.

2 Ronco C, Bellomo R: Acute renal failure and multiple organ dysfunction in the ICU: from renal replacement therapy (RRT) to multiple organ support therapy (MOST). Int J Artif Organs 2002;25:733–747.

3 Hirasawa H, Oda S, Matsuda K: Continuous hemodiafiltration with cytokine-adsorbing hemofilter in the treatment of severe sepsis and septic shock. Contrib Nephrol 2007;156:365–370.

4 VA/NIH Acute Renal Failure Trial Network, Palevsky PM, Zhang JH, O'Connor TZ, Chertow GM, Crowley ST, Choudhury D, Finkel K, Kellum JA, Paganini E, Schein RM, Smith MW, Swanson KM, Thompson BT, Vijayan A, Watnick S, Star RA, Peduzzi P: Intensity of renal support in critically ill patients with acute kidney injury. N Engl J Med 2008;359:7–20.

5 Marshall MR, Golper TA, Shaver MJ, Alam MG, Chatoth DK: Sustained low-efficiency dialysis for critically ill patients requiring renal replacement therapy. Kidney Int 2001;60:777–785.

6 Tolwani AJ, Wheeler TS, Wille KM: Sustained low-efficiency dialysis. Contrib Nephrol 2007;156:320–324.

7 Kumar VA, Craig M, Depner TA, Yeun JY: Extended daily dialysis: a new approach to renal replacement for acute renal failure in the intensive care unit. Am J Kidney Dis 2000;36:294–300.

8 Lonnemann G, Floege J, Kliem V, Brunkhorst R, Koch KM: Extended daily veno-venous high-flux haemodialysis in patients with acute renal failure and multiple organ dysfunction syndrome using a single path batch dialysis system. Nephrol Dial Transplant 2000;15:1189–1193.

9 Henderson L, Besarab A, Michaels A, Bluemle LW: Blood purification by ultrafiltration and fluid replacement (dialfiltration). Trans Am Soc Artif Intern Organs 1967;13:216–226.

10 Leber HW, Wizemann V, Goubeaud G, Rawer P, Schutterle G: Hemodiafiltration: a new alternative to hemofiltration and conventional hemodialysis. Artifi Organs 1978;2(suppl):408–411.

11 Kramer P, Wigger W, Rieger J, Matthaei D, Scheler F: A new and simple method for treatment of over-hydrated patients resistant to diuretics (in German). Klin Wochenschr 1977;55:1121–1122.

12 Ronco C, Bellomo R, Homel P, Brendolan A, Dan M, Piccinni P, La Greca G: Effects of different doses in continuous veno-venous haemofiltration on outcomes of acute renal failure: a prospective randomised trial. Lancet 2000;356:26–30.

13 Mineshima M, Ishimori I, Sakiyama R: Validity of internal filtration-enhanced hemodialysis as a new hemodiafiltration therapy. Blood Purif 2009;27:33–37.

14 Locatelli F, Di Filippo S, Manzoni C: Removal of small and middle molecules by convective techniques. Nephrol Dial Transplant 2000;15(suppl 2):37–44.
15 Parzer S, Balcke P, Mannhalter C: Plasma protein adsorption to hemodialysis membranes: studies in an in vitro model. J Biomed Mater Res 1993;27:455–463.
16 Nakamura M, Oda S, Sadahiro T, Hirayama Y, Tateishi Y, Abe R, Hirasawa H: The role of hypercytokinemia in the pathophysiology of tumor lysis syndrome (TLS) and the treatment with continuous hemodiafiltration using a polymethylmethacrylate membrane hemofilter (PMMA-CHDF). Transfus Apher Sci 2009;40:41–47.

Hideki Kawanishi, MD
Tsuchiya General Hospital,
3–30 Nakajima-cho, Naka-ku
Hiroshima 730-8655 (Japan)
Tel. +81 82 243 9191, Fax +81 82 241 1865, E-Mail h-kawanishi@tsuchiya-hp.jp

Suzuki H, Hirasawa H (eds): Acute Blood Purification.
Contrib Nephrol. Basel, Karger, 2010, vol 166, pp 21–30

Indications for Blood Purification in Critical Care

Hiroyuki Hirasawa

Department of Emergency and Critical Care Medicine, Chiba University Graduate School of Medicine, Chiba, Japan

Abstract

Blood purification in critical care can perform 2 main functions: as an artificial support for failing organs (such as artificial kidney or liver support) and as a remover of causative humoral mediators of critical illness (such as severe sepsis and acute respiratory distress syndrome). As an artificial kidney, continuous blood purification (such as continuous hemofiltration and continuous hemodiafiltration, CHDF) is widely applied in intensive care units. The intensity of renal replacement therapy, however, has been reported to have no impact upon the efficacy of the blood purification in terms of clinical outcome. Concerning blood purification and the removal of causative humoral mediators of critical illness, CHDF using a hemofilter made from polymethylmethacrylate membrane is reported to be very effective in the treatment of severe sepsis and septic shock, even in septic patients without renal dysfunction. Thus, in Japan, CHDF with a polymethylmethacrylate membrane is now widely applied for non-renal indications, not only for patients with sepsis but also patients with cytokine-induced critical illness (such as acute respiratory distress syndrome and severe acute pancreatitis), even when those patients do not present with renal dysfunction. However, our understanding of the pathophysiology of sepsis has changed since the concept of pattern recognition receptors and pathogen-associated molecular patterns was introduced. According to this, CHDF with a cytokine-adsorbing polymethylmethacrylate membrane hemofilter is preferable and more effective than direct hemoperfusion with an endotoxin-adsorbing polymyxin-B immobilized column in the treatment of sepsis and septic shock. Blood purification in critical care is gaining popularity, and is widely for both renal and non-renal indications.

Blood Purification in Intensive Care Units

In intensive care units (ICU), where critically ill patients often present with acute organ dysfunction (such as acute respiratory failure and acute renal failure), many kinds of artificial organ supports are utilized when the degree of organ dysfunction has deteriorated beyond the point necessary to sustain life. When we studied the frequency of use of the various artificial supports on critically ill patients in our ICU, we found that the many techniques of blood purification are the second most frequently applied form of artificial support, next to the ventilator. Thus, blood purification is now an invaluable modality in critical care.

Conventionally, many kinds of blood purification are applied in ICU as artificial support for failing organs (such as an artificial kidney for patients with acute renal failure [1] and artificial liver support for patients with acute liver failure [2]). Moreover, it has been suggested that continuous hemofiltration (CHF) or continuous hemodiafiltration (CHDF) can remove many kind of causative humoral mediators of various critical illness [3]. Taking this into consideration, CHF/CHDF has been applied to remove such causative humoral mediators under 'non-renal indications', even when critically ill patients presented without renal dysfunction.

In our ICU, patients with tumor lysis syndrome [4], hemophagocytic syndrome [5], severe acute pancreatitis [6] and intracranial hypertension [7] have successfully been treated with blood purification under non-renal indications.

The Surviving Sepsis Campaign guidelines [8] were revised in 2008, accepted worldwide, and applied to patients with severe sepsis and septic shock, resulting in improved survival [9]. The description of blood purification in the Surviving Sepsis Campaign guidelines is, however, as follows [8]:

> We suggest that continuous renal replacement therapies and intermittent hemodialysis are equivalent in patients with severe sepsis and acute renal failure (grade 2B). We suggest that use of continuous therapies to facilitate management of fluid balance in hemodynamically unstable septic patients (grade 2D).

Consequently, there is no description or recommendation concerning blood purification aimed at removing causative humoral mediators of sepsis.

Figure 1 shows the details of the blood purifications performed in our ICU over the past 18 years. Blood purification was given to 18.7% of all ICU patients. The most frequently performed blood purification was CHDF (93.8 % of these patients), which was followed by plasma exchange (16.6%). Therefore, we suggest that if CHDF and plasma exchange are available in the ICU, almost all ICU patients who need some blood purification therapy can be treated effectively. Taking these data into the consideration, we will briefly review the indications for blood purification in ICU.

Indications for Blood Purification in ICU

As mentioned previously, there are 2 kinds of indications of blood purification in critical care. The first is when artificial support for failing organs is

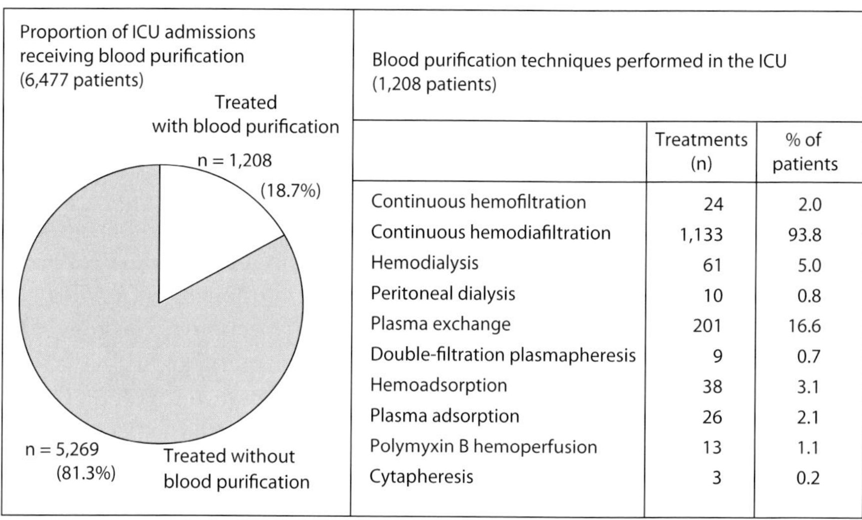

Blood purification techniques performed in the ICU (1,208 patients)

	Treatments (n)	% of patients
Continuous hemofiltration	24	2.0
Continuous hemodiafiltration	1,133	93.8
Hemodialysis	61	5.0
Peritoneal dialysis	10	0.8
Plasma exchange	201	16.6
Double-filtration plasmapheresis	9	0.7
Hemoadsorption	38	3.1
Plasma adsorption	26	2.1
Polymyxin B hemoperfusion	13	1.1
Cytapheresis	3	0.2

Fig. 1. Blood purification treatments performed at the ICU in Chiba University Hospital (1981–2008).

required (acute renal and liver failures being the most common presentations), and the other is when the removal of causative humoral mediators of critical illness (such as sepsis [10] and severe acute pancreatitis [6]) is necessary.

Regarding the use of blood purification as an artificial organ, especially as an artificial kidney, there is some controversy. The main issues concern intermittent versus continuous blood purification and its intensity. As mentioned in the Surviving Sepsis Campaign guidelines, it is generally accepted that continuous renal replacement therapies and intermittent hemodialysis are equally effective in patients with severe sepsis and acute renal failure [8, 9]. Furthermore, regarding the efficacy of increased intensity of blood purification when used as artificial kidney, it is generally agreeed that increased intensity of renal support in critically ill patients with acute renal failure is not associated with improved mortality and improved renal recovery [11–13]. This is despite the fact that the well-known article by Ronco et al. [14] indicated a beneficial effect of increased intensity of renal support on survival of the patients with acute renal failure. The efficacy of various kinds of blood purification on patients with acute liver failure is discussed in other chapters.

Discussions on using blood purification in ICU in the removal of humoral mediators have centered around the removal of cytokine and causative humoral mediators of sepsis, and will be discussed in the following section.

Blood Purification for Hypercytokinemia and Sepsis

Hypercytokinemia in Critical Illness and Sepsis

It is now widely accepted through the concept of the 'cytokine theory of disease' that hypercytokinemia plays a pivotal role in many critical illness treated in ICU, such as sepsis. Lowry [15] stated in his recent review article: 'In many respects, the scientific accomplishments of the Shires era in the Department of Surgery at Cornell can be defined as an underpinning for the cytokine theory of disease.' However, since then, our understanding of the pathophysiology of sepsis has been changed and elucidated. Until recently, endotoxin is thought to be the most important bacterial products that induce hypercytokinemia release in sepsis. However, after introducing the concept of pattern recognition receptors (PRR), alarmin, and pathogen-associated molecular patterns (PAMP), it is now considered that many microbial products other than endotoxin (such as flagellin, lipoprotein and peptidoglycan) can also be a ligand to PRR promoting the release of cytokines [16, 17]. Furthermore, now it is widely accepted that alarmin or DAMP (endogenous danger-associated molecular patterns) – such as heat shock protein, HMGB-1 (high motility group box 1) and even necrotic cells – can be the ligands to PRRs resulting in the release of cytokines. Therefore, the importance of endotoxin as the ligands initiating cytokine release in sepsis was very much decreased [16, 17]. This suggests that we should consider the removal of cytokine as a therapeutic strategy in not only endotoxemic patients, but also in all septic patients and in patients with many other critical illnesses (such as acute respiratory distress syndrome and severe acute pancreatitis) where hypercytokinemia is thought to play an important pathophysiological role.

Blood Purification for Hypercytokinemia

Blood purification for hypercytokinemia has been extensively discussed, especially in hypercytokinemia sepsis patients [10, 18]. Since the study of Ronco et al. [15], it has become generally accepted that high-volume hemofiltration should be performed to remove cytokines from the blood stream [19, 20]. Concerning the mechanism of cytokine removal with hemofiltration, 3 possible mechanisms have been proposed: peak concentration theory [3], threshold modulation theory [10] and mediator delivery theory [21]. Further investigation is needed to elucidate which theory is most important in this aspect.

Our group repeatedly reported that CHDF using a cytokine-adsorbing hemofilter made from polymethylmethacrylate membrane (PMMA-CHDF) can continuously and effectively remove many kinds of pro- and anti-inflammatory cytokines and decrease the blood levels of those cytokines [18]. Furthermore, we also reported that PMMA-CHDF is very effective in the treatment of severe sepsis and septic shock [22, 23], even when those patients presented without renal dysfunction. It was also reported that this efficacy could not be achieved with CHDF using a hemofilter made from polyacrylonitrile membrane [23]. We further reported that the reason for the better efficacy of PMMA-CHDF on cytokine removal compared to CHDF using a hemofilter made from other membrane material is because the main mechanism of cytokine removal in PMMA-CHDF is through the cytokine adsorption to the hemofilter and this kind of cytokine adsorption is very membrane-specific; hemofilters made from materials other than PMMA do not have enough cytokine-adsorbing capacity [18]. When we use a PMMA hemofilter, we do not need to perform high-volume CHF or high-volume CHDF (which are sometimes troublesome and expensive) to remove cytokines with continuous blood purification [18]. We have seen many cases of hypercytokinemia-induced critical illnesses (such as tumor lysis syndrome [3], hemophagocytic syndrome [4] and severe acute pancreatitis [5]) treated effectively with PMMA-CHDF. Indications for treatment with PMMA-CHDF include:

- severe sepsis, septic shock;
- acute respiratory distress syndrome;
- postoperative persistent hypercytokinemia;
- severe acute pancreatitis;
- congestive heart failure, cardiogenic shock;
- hematological disorders, malignant tumor;
- thrombotic microangiopathy;
- trauma, hemorrhagic shock;
- post-cardiopulmonary arrest;
- others.

PMMA-CHDF or PMX-DHP for Sepsis

In Japan, direct hemoperfusion with the polymyxin-B immobilized column (PMX-DHP) is frequently given to patients with sepsis, aiming to remove endotoxin that used to be thought of as the most important pathogenic factor in sepsis. However, as mentioned earlier in this paper, since the introduction of the concept of PRRs, alarmin and PAMP, our understanding of the pathophysiology of sepsis has changed greatly. Now it is reasonable to say that endotoxin is at most just one of the PAMP. Recently, a randomized controlled trial of PMX-DHP on patients with septic shock of abdominal origin (EUPHAS) was published and reported positive results [24]. However, there are some questions regarding the EUPHAS study. First of all, the survival of the control group being treated conventionally without PMX-DHP was only 47%, which is low compared to generally reported survival of septic shock patients. The survival of the PMX-DHP group was 68%, which is worse than the 70% survival of patients with severe sepsis and septic shock treated according to the Surviving Sepsis Campaign guidelines without PMX-DHP [9]. The survival of the patients with septic shock treated with PMMA-CHDF reported by Nakada et al. [22] was 79.1%. Furthermore, in the EUPHAS study, the enrollment criterion was solely abdominal origin

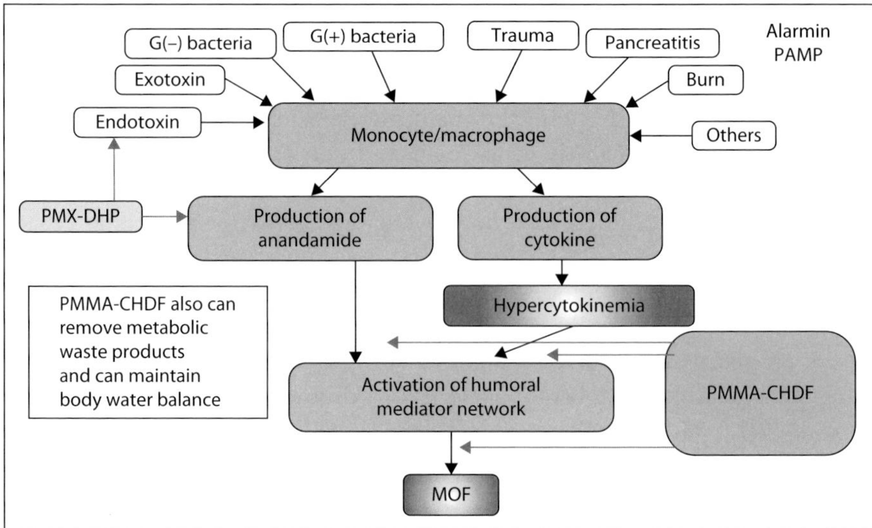

Fig. 2. Efficacy of PMX-DHP and PMMA-CHDF for the humoral mediator network. MOF = Multiple Organ Failure.

septic shock, which is thought to be one of most easily treatable types of sepsis. Therefore, the results of this study cannot be generalized to entire population of septic shock patients. Therefore, as shown in figure 2, we suggest that PMMA-CHDF is a more effective and widely applicable form of blood purification for severe sepsis and septic shock compared to PMX-DHP [25].

Taking the new concepts regarding the pathophysiology of severe sepsis/septic shock into consideration, we propose a new therapeutic strategy of severe sepsis/septic shock featuring hypercytokinemia and countermeasures against hypercytokinemia using PMMA-CHDF (fig. 3).

Future Perspectives

There are several critical illnesses and pathological conditions other than those already mentioned which could be treated with blood purification in future. One of them is post-cardiac arrest syndrome. The main pathophysiological factor of post-cardiac arrest syndrome is hypercytokinemia [26], and Laurent et al. [27] reported that high-volume CHF could improve the overall prognosis after resuscitation.

It has recently been reported that the severity of hypercytokinemia following many kinds of insult (including infection and trauma) is affected by the presence or absence of cytokine-related genetic polymorphisms, and that the

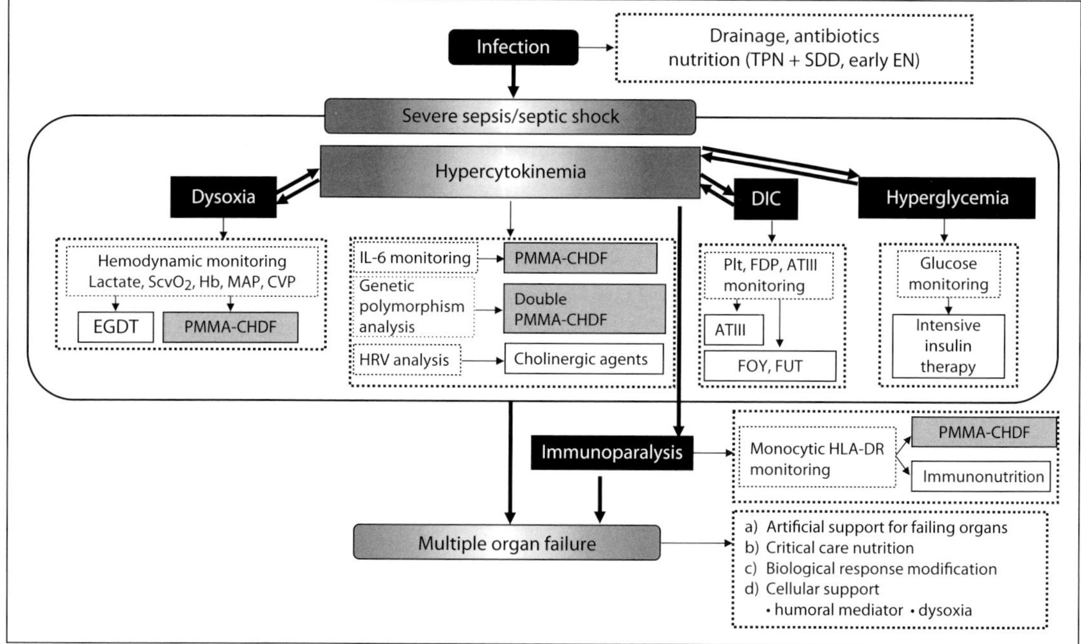

Fig. 3. Therapeutic strategy for severe sepsis/septic shock according to hypercytokinemia theory. TPN = Total Parenteral Nutrition; SDD = Selective Digestive Decontamination; EN = Enteral Nutrition; DIC = Disseminated Intravascular Coagulation; MAP = Mean Arterial Pressure; CVP = Central Venous Pressure; EGDT = Early Goal-Directed Therapy; HRV = Heart Rate Valiability; Plt = Platelet; FDP = Fibrinogen Degradation Product; ATIII = angiotensin III; FOY = Trade name of Gabexate Mesilate; FUT = Nafamostat Mesilate; HLA-DR = HLA-DR antigen.

patients with a cytokine-related genetic polymorphism presented severer hypercytokinemia compared to the patients without a cytokine-related genetic polymorphism [28]. It was also reported that PMMA-CHDF is less effective as a cytokine remover in patients with a cytokine-related genetic polymorphism [28]. Therefore, in the future, some kind of tailor-made medicine according to the cytokine-related genetic polymorphism should be considered when we apply PMMA-CHDF as a cytokine modulator [28].

In this short review, we have focused mainly on the efficacy of the removal of pro-inflammatory cytokines in the treatment of severe sepsis and septic shock. However, our preliminary report indicates that removal of the anti-inflammatory cytokine interleukin-10 with PMMA-CHDF results in recovery from immunoparalysis.

Concerning the use of blood purification for humoral mediator modulation, coupled plasma filtration adsorption therapy [29] and direct hemoperfusion with a cytokine-adsorbing column [30] have been investigated, and the results of the multicenter trials of these new modalities are awaited.

Indications for blood purification have been greatly expanded in critical care. Its use in removing causative humoral mediators of critical illness has attracted a great deal of attention and is gaining popularity. Currently, blood purification is of the most important modalities in ICU, along with nutritional support (with total parenteral nutrition or tube feeding) and ventilatory support (using a ventilator).

References

1 Hirayama Y, Hirasawa H, Oda S, Shiga H, Nakanishi K, Matsuda K, Nakamura M, Hirano T, Moriguchi T, Watanabe E, Nitta M, Abe R, Nakada T: The change in renal replacement therapy on acute renal failure in a general intensive care unit in a university hospital and its clinical efficacy: a Japanese experience. Ther Apher 2003;7:475–482.

2 Yokoi T, Oda S, Shiga H, Matsuda K, Sadahiro T, Nakamura M, Hirasawa H: Efficacy of high-flow dialysate continuous hemodiafiltration in the treatment of fulminant hepatic failure. Transfus Apher Sci 2009; 40:61–70.

3 Ronco C, Tetta C, Mariano F, Wratten ML, Bonello M, Bordoni V, Cardona X, Inguaggiato P, Pilott L, d'Intini V, Bellomo R: Interpreting the mechanisms of continuous renal replacement therapy in sepsis: the peak concentration hypothesis. Artif Organs 2003;27:792–801.

4 Nakamura M, Oda S, Sadahiro T, Hirayama Y, Tateishi Y, Abe R, Hirasawa H: The role of hypercytokinemia in the pathophysiology of tumor lysis syndrome (TLS) and the treatment with continuous hemodiafiltration using a polymethyl methacrylate membrane hemofilter (PMMA-CHDF). Transfus Apher Sci 2009;40:41–47.

5 Tateishi Y, Oda S, Sadahiro T, Nakamura M, Hirayama Y, Abe R, Hirasawa H: Continuous hemodiafiltration in the treatment of reactive hemophagocytic syndrome refractory to medical therapy. Transfus Apher Sci 2009; 40:33–40.

6 Oda S, Hirasawa H, Shiga H, Matsuda K, Nakamura M, Watanabe E, Moriguchi T: Management of intra-abdominal hypertension in patients with severe acute pancreatitis with continuous hemodiafiltration using a polymethyl methacrylate membrane hemofilter. Ther Apher Dial 2005;9:355–361.

7 Nakanishi K, Hirasawa H, Oda S, Shiga H, Matsuda K, Nakamura M, Hirano T, Hirayama Y, Moriguchi T, Watanabe E, Nitta M: Intracranial pressure monitoring in patients with fulminant hepatic failure treated with plasma exchange and continuous hemodiafiltration. Blood Purif 2005;23: 113–118.

8 Dellinger RP, Levy MM, Carlet JM, Bion J, Parker MM, Jaeschke R, Reinhart K, Angus DC, Brun-Buisson C, Beale R, Calandra T, Dhainaut J-F, Gerlach H, Harvery M, Marini JJ, Marshall J, Ranieri M, Ramsay G, Sevransky J, Thompson T, Townsend S, Vender JS, Zimmerman JL, Vincent J-L, International Surviving Sepsis Campaign Guidelines Committee: Surviving Sepsis Campaign: international guidelines for management of severe sepsis and septic shock: 2008. Crit Care Med 2008;36:296–327.

9 Levy MM, Dellinger RP, Townsend SR, LInde-Zwirble WT, Marshall JC, Bion J, Schorr C, Artigas A, Ramsay G, Beale R, Parker MM, Gerlach H, Reinhart K, Silva E, Harvey M, Regan S, Angus DC, Surviving Sepsis Campaign: The Surviving Sepsis Campaign: results of an international guideline-based performance improvement program targeting severe sepsis. Crit Care Med 2010;30:367–374.

10 Honore PM, Joannes-Boyau O, Boer W, Colln V: High-volume hemofiltration in sepsis and SIRS: current concepts and future prospects. Blood Purif 2009;28:1–11.
11 Pannu N, Klarenbach S, Wiebe N, Manns B, Tonelli M, Alberta Kidney Disease Network: Renal replacement therapy in patients with acute renal failure. JAMA 2008;299:793–805.
12 RENAL Study Investigators, Bellomo R, Cass A, Cole L, Finfer S, Gallagher M, Goldsmith D, Myburgh J, Norton R, Scheinkestel C: Design and challenges of the randomized evaluation of normal versus augmented level replacement therapy (RENAL) trial: high-dose versus standard-dose hemofiltration in acute renal failure. Blood Purif 2008;26:407–416.
13 VA/NIH Acute Renal Failure Trial Network: Intensity of renal support in critically ill patients with acute renal failure. N Engl J Med 2008;359:7–20.
14 Ronco C, Bellomo R, Homel P, Brendolan A, Dan M, Piccinni P, La Greca G: Effects of different doses in continuous veno-venous haemofiltration on outcomes of acute renal failure: a prospective randomized trial. Lancet 2000;355:26–30.
15 Lowry SF: The evolution of an inflammatory response. Surg Infect 2009;10:419–425.
16 Harris HE, Raucci A: Alarmin(g) news about danger: workshop on innate danger signals and HMGB1. EMBO Rep 2006;7:774–778.
17 Rittirsch D, Flierl MA, Ward PA: Harmful molecular mechanisms in sepsis. Nat Rev Immuno 2008;8:776–787.
18 Nakada T, Hirasawa H, Oda S, Shiga H, Matsuda K: Blood purification for hypercytokinemia. Transfus Apher Sci 2006;35:253–264.
19 Ronco C, Kellum JA, Bellomo R, House AA: Potential interventions in sepsis-related acute kidney injury. Clin J Am Soc Nephrol 2008; 3:531–544.
20 Payen D, Mateo J, Cavaillon JM, Fraise F, Floriot C, Vicaut E, Hemofiltration and Sepsis Group of the College National de Reanimation et de Medicine d'Urgence des Hopitaux extra-Universitaires: Impact of continuous venovenous hemofiltration on organ failure during the early phase of severe sepsis: a randomized controlled trial. Crit Care Med 2009;37:803–810.
21 Di Carlo JV, Alexander SR: Hemofiltration for cytokine-driven illness: the mediator delivery hypothesis. Internat J Artif Organs 2005;28:777–786.
22 Nakada T, Oda S, Matsuda K, Sadahiro T, Nakamura M, Abe R: Continuous hemodiafiltration with PMMA hemofilter in the treatment of patients with septic shock. Mol Med 2008;14:257–263.
23 Matsuda K, Moriguchi T, Harii N, Goto J: Comparison of efficacy between continuous hemodiafiltration with a PMMA membrane hemofilter and a PAN membrane hemofilter in the treatment of a patient with septic acute renal failure. Transfus Apher Sci 2009;40:49–53.
24 Cruz DN, Antonelli M, Funagalli R, Foltran F, Brienza N, Donati A, Malacangi V, Petrini F, Volta G, Pallavicini FMB, Rottoli F, Giunta F, Ronco C: Early use of polymyxin B hemoperfusion in abdominal septic shock: the EUPHAS randomized controlled trial. JAMA 2009;301:2445–2452.
25 Hirasawa H, Oda S, Shiga H, Matsuda K: Endotoxin adsorption or hemodiafiltration in the treatment of multiple organ failure. Curr Opin Crit Care 2000;6:421–425.
26 Nolan JP, Neumar RW, Adrie C, Aibiki M, Berg RA, Bottiger BW, Callaway C, Clark RSB, Geocadin RG, Jauch EC, Kern KB, Laurent I, Longstreth WT, Merchant RM, Morley P, Morrison LJ, Nadkarni V, Peberdy MA, Rivers EP, Rodriguez-Nunez A, Sellke FW, Spaulding C, Sunde K, Hoek TV: Post-cardiac arrest syndrome: Epidemiology, pathophysiology, treatment, and prognostication. A scientific statement from the International Liaison Committee on Resuscitation: The American Heart Association Emergency Cardiovascular Care Committee; The Council on cardiopulmonary, perioperative, and critical care: The council on clinical cardiology: the Council on Stroke. Resuscitation 2008;79:350–379.
27 Laurent I, Adrie C, Vinsonneau C, Cariou A, Chiche J-D, Ohanessian A, Spaulding C, Carli P, Dhainaut J-F, Monchi M: High-volume hemofiltration after out-of-hospital cardiac arrest. J Am Coll Cardiol 2005;46: 432–437.

28 Watanabe E, Hirasawa H, Oda S, Matsuda K, Hatano M, Tokuhisa T: Extremely high interleukin-6 blood levels and outcome in the critically ill are associated with tumor necrosis factor- and interleukn-1-related gene polymorphism. Crit Care Med 2005;33:89–97.
29 Ronco C, Brendolan A, d'intini V, Ricci Z, Wratten ML, Bellomo R: Coupled plasma filtration adsorption: rationale, technical development and early clinical experience. Blood Purif 2003;21:409–416.
30 Kobe Y, Oda S, Matsuda K, Nakamura M, Hirasawa H: Direct hemoperfusion with a cytokine-adsorbing device for the treatment of persistent or severe hypercytokinemia: a pilot study. Blood Purif 2007;25:446–453.

Hiroyuki Hirasawa, MD, PhD
Department of Emergency and Critical Care Medicine, Chiba University Graduate School of Medicine
1–8–1 Inohana, Chuo
Chiba 260-8677 (Japan)
Tel. +81 43 226 2341, Fax +81 43 226 2173, E-Mail hhirasawa@faculty.chiba-u.jp

Suzuki H, Hirasawa H (eds): Acute Blood Purification.
Contrib Nephrol. Basel, Karger, 2010, vol 166, pp 31–39

Acute Kidney Injury of Non-Septic Origin Requiring Dialysis Therapy

Hiromichi Suzuki[a,b] · Yoshihiko Kanno[a] · Isao Tsukamoto[b] · Youhei Tsuchiya[b] · Soichi Sugahara[a]

[a]Department of Nephrology, Saitama Medical University, and [b]Division of Dialysis Center, Saitama International Medical Center, Saitama, Japan

Abstract

Acute kidney injury (AKI) requiring dialysis occurs frequently, and its pathogenesis involves multiple pathways within which hemodynamic, inflammatory and nephrotoxic factors overlap. Several studies have tried to assess the risk factors leading to AKI, and found, among other factors, that preoperative renal dysfunction is important. Currently, it is uncertain when dialysis therapy should start. However, AKI after cardiac surgery should be treated early by continuous hemodialysis.

Two major causes of acute kidney injury (AKI) requiring renal replacement therapy are sepsis and cardiovascular surgery. AKI is a frequent and severe complication after cardiovascular surgery. The need for dialysis therapy occurs in approximately 1% of patients [1–3], and the development of renal failure is associated with high mortality.

Numerous variables have been identified as risk factors for the development of AKI in this setting, including age, male gender, diabetes, New York Heart Association class, peripheral vascular disease, chronic obstructive pulmonary disease, chronic kidney disease, hypotension during surgery, use of vasoactive drugs, duration of cardiopulmonary bypass, and aortic clamping [4, 5].

Pathogenesis of AKI in Cardiovascular Surgery

The pathogenesis of AKI due to post-cardiac surgery is heterogeneous; however, the 2 major mechanisms by which the kidney sustains injury are ischemic and direct nephrotoxic effects. Although the kidneys receive more blood flow than any other major organ, they are the most vulnerable to ischemic injury. Renal blood flow is regulated by several systems, such as the renin-angiotensin system and nitric oxide. Inadequate renal perfusion during cardiovascular surgery is the main source of ischemia-perfusion injury.

Predictive Factors for the Risk of Postoperative Dialysis

The average age of patients referred for cardiovascular surgery is becoming older and they also present with comorbid conditions [6, 7]. A means of estimating a patient's individual risk of requiring postoperative dialysis (a more important adverse event from patient's and physician's point of view) based on multiple preoperative clinical features would facilitate internal clinical decision-making and patient consultations.

The Society of Thoracic Surgeons' National Cardiac Surgery Database provides an opportunity to study a nationwide sample of patients undergoing cardiac surgery in a diverse cross-section of US hospitals [8, 9]. Using these data, Mehta et al. [10] identified preoperative clinical variables and operative types associated with the need for postoperative dialysis after coronary artery bypass graft (CABG) surgery and/or valvular heart surgery. Using these identifiers, they sought to create an efficient bedside tool to estimate the need for dialysis after cardiac surgery. They then examined the degree to which this tool could estimate risk among diverse patient subgroups and among those undergoing different cardiac surgical procedures (including isolated CABG, isolated aortic valve or isolated mitral valve surgery, CABG plus aortic valve surgery, and CABG plus mitral valve surgery).

Previous studies have demonstrated that a postoperative need for renal dialysis occurs in 1.1–3.0% of cardiac surgery cohorts [11–13]. Importantly, these studies indicate that the minority of patients who require dialysis after cardiac surgery have significantly longer in-hospital stays and extremely high mortality, with 63–100% of patients dying before leaving the hospital [11, 12]. The contemporary data of Mehtas et al. [10] suggest that the need for dialysis after cardiac surgery in the community at large is on the lower side of prior estimates (1.4%), despite the fact that surgery is increasingly being performed on elderly high-risk patients. Nevertheless, postoperative dialysis remains associated with significant morbidity and mortality.

These risks of postoperative dialysis are clearly related to preoperative kidney function. These results support and extend the work of others. One prior study [12] of renal function and outcome suggested that patients with a serum creatinine level >2.6 mg/dl are at extreme risk of dialysis dependency after CABG, and that perhaps alternative options for coronary management should be strongly considered. In contrast, the data of Mehtas et al. [10] suggest that this relationship is continuous and that risk of postoperative dialysis parallels the increase in serum creatinine. However, preoperative serum creatinine alone is a relatively poor discriminator of those who will require dialysis. For example, although the average preoperative serum creatinine of those requiring postoperative dialysis is 1.6 mg/dl, up to three quarters of those with preoperative serum creatinine levels of 3.5 mg/dl will not require postoperative dialysis.

Most previous studies have assessed the association between renal function and outcome after cardiac surgery by dichotomizing renal function, using a plasma creatinine level between 1.4 and 1.7 mg/dl as a cutoff point to diagnose mild renal dysfunction [14–16]. By using estimated Cl_{Cr} and plasma creatinine levels as continuous variables in multiple regression analyses, this study better defines and quantifies the relationship between those measures of renal function and outcome after cardiac surgery. After adjusting for confounding variables, the odds of developing renal failure requiring dialysis, dying, or having major morbidity after cardiac surgery increase by 52, 27 and 18% for each 10 ml/min/1.73 m^2 decrement in estimated Cl_{Cr}, respectively. Similarly, for each 0.2 mg/dl increment in plasma creatinine level, the risk-adjusted odds of having the same adverse outcomes increase by 20, 8 and 13%, respectively.

In the study of Wang et al. [17], there were 4,603 patients with normal plasma creatinine levels in the study. The rates of postoperative renal failure requiring dialysis, mortality and major morbidity in those patients were 0.8, 2.2 and 17.9%, respectively. Multiple regression analyses repeated only in those patients with normal plasma creatinine levels showed that estimated Cl_{Cr} remained a significant risk predictor of all outcomes. Each 10 ml/min/1.73 m^2 decrement in estimated Cl_{Cr} was associated with ORs of 1.59 (95% CI 1.27–1.98) for renal failure requiring dialysis, 1.34 (95% CI 1.19–1.51) for mortality, and 1.12 (95% CI 1.08–1.17) for major morbidity. Emergency surgery, combined/complex surgical procedures, congestive heart failure and age also remained significant risk factors for all studied outcomes in patients with normal plasma creatinine levels; however, plasma creatinine level was not a significant predictor for any outcome in that particular population. With each 0.2 mg/dl increase in plasma creatinine level, the ORs were 1.02 (95% CI 0.75–1.41) for mortality and 1.08 (95% CI 0.96–1.22) for major morbidity.

Intraoperative Events

Numerous procedural factors have been associated with AKI after cardiopulmonary bypass (CPB) surgery. Clearly, this is a vulnerable period with hemodynamic alterations and activation of the immune system. The data presented previously, which demonstrate renal protection with off-pump CAB surgery, add to the evidence that CPB itself may be injurious to the kidney. The ultimate goal of CPG is to maintain regional perfusion at a level that supports optimal cellular and organ function. Any decrease in renal perfusion during CPG, depending on its magnitude and duration, can lead to significant cellular injury. Generally, CPB flow rates of 1.8–2.2 $l/min/m^2$ are recommended along with a mean perfusion pressure of 50–70 mm Hg. Little is known about the effect of this flow rate and perfusion pressure on regional renal blood flow and local oxygen delivery rates [18]. Most studies on the autoregulation of regional blood flow during CPB have focused on cerebral circulation, and demonstrated preserved cerebral autoregulation with these parameters.

Postoperative Events

Postoperative events – such as the need for vasoactive agents, hemodynamic instability, exposure to nephrotoxic medications, volume depletion and sepsis/SIRS – are all critical events that can lead to kidney injury. Perhaps the most critical factor is postoperative cardiac performance and the need for either inotropic or mechanical support for left ventricular dysfunction, the risk of significant renal injury becomes very high as the vulnerable kidney is subjected to marginal perfusion pressures and worsening of the ischemic injury can occur.

Wijeysundera et al. [19] clearly demonstrated that a predictive index with a simplified scoring system could use readily available and clinically sensitive preoperative information to provide accurate prognostic information on RRT after cardiac surgery. The simplified renal index warrants comparison against the few previous predictive indices for RRT after cardiac surgery [10, 20]. Our index demonstrated improved ease of use because of a smaller number of component variables and a simpler scoring system, yet showed similar prognostic accuracy. In this respect, the simplified renal index mirrors the evolution of preoperative cardiac risk indices, where the newer Revised Cardiac Risk Index has comparable accuracy, but improved usability, compared with the older indices [21, 22].

Beginning Dialysis Therapy

Sugahara and Suzuki [23] tried to determine whether starting hemodialysis therapy early was effective in improving the prognosis of patients with AKI following CABG.

Of 486 patients who underwent cardiac surgery at Saitama Medical School Hospital during the period from January 1, 1995 to December 31, 1997, 40 patients had AKI following CABG and were recruited into the study. Patients were entered into the study when hourly urinary output became ≤30 ml/h and serum creatinine increased at the rate of ≥0.5 mg/dl/day. All patients were divided randomly into 2 groups: an 'early start' treatment group that consisted of 14 patients who received dialysis when hourly urinary output became <30 ml/h for 3 consecutive hours (or daily urinary output was approximately ≤750 ml) and a 'conventional start' treatment group that consisted of 14 patients who received dialysis when hourly urinary output became <20 ml/h for 2 consecutive hours (or daily urinary output was approximately ≤500 ml). Patients were accessed through double-lumen catheters (Vas-cath, Medicon, Chicago, Ill., USA), which were inserted into the right or left femoral vein and connected to a continuous hemodialyzer (KM8600, Kurary, Tokyo, Japan). An anticoagulant, nafamostat mesilate (Futhan, Torii Pharmaceutical, Tokyo, Japan), was used at 30 IU/h. Dialysis started under conditions of water elimination rate of 60 ml/h and a dialysate (HF Solita, Shimizu Pharmaceutical, Shimizu, Japan) flow rate of 1 l/h. The dialyzers used in this study were Panflow APF-S (Asahi Medical, Tokyo, Japan) and Hemofeel SH (Toray Medical, Tokyo, Japan).

Comparison of Data at the Start of Continuous Hemodialysis Therapy

Urinary Output
Urinary output of 2 groups at the start of continuous hemodialysis was 29 ± 1 ml/h for the early start group and 18 ± 1 ml/h for the conventional start group ($p < 0.05$).

Levels of Serum Creatinine
The mean serum creatinine level was not significantly different between the 2 groups, although it was significantly elevated compared to the preoperative value in both groups.

Blood Pressure
There was no significant difference between the 2 groups in either the mean systolic or the diastolic blood pressures.

Changes in Blood Pressure (Fig. 1)
In both groups, blood pressure decreased in the first 3 days after the start of continuous hemodialysis and tended to increase thereafter.

Changes in Urinary Output (Fig. 2)
There was no change in urinary output in the first 3 days after the start of continuous hemodialysis in the early start group, whereas urinary output decreased markedly in the conventional start group. On the sixth day and thereafter, urinary output began to increase gradually in the early start group and by the eighth day there were significant differences in urinary output from those at the start of dialysis ($p < 0.05$). In the conventional start group, although urinary output showed a tendency to increase, there was no statistical significance, likely due to the small number of subjects.

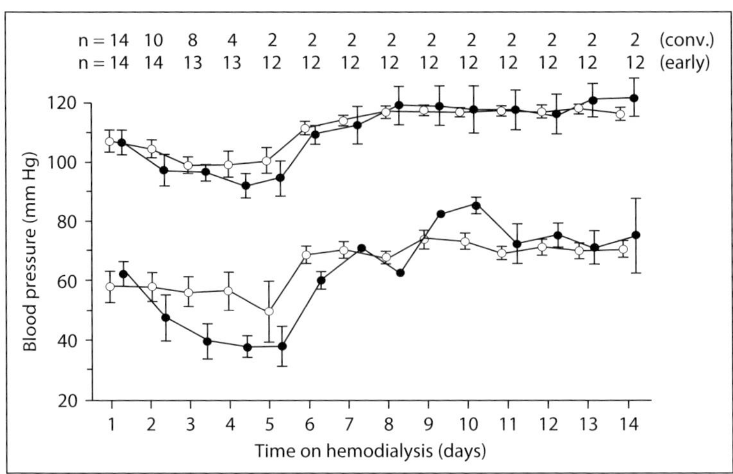

Fig. 1. Changes in systolic and diastolic blood pressure after the initiation of dialysis therapy. There was no significant difference between 2 groups. ○ = Early intervention; ● = conventional intervention. Values represent means ± SD.

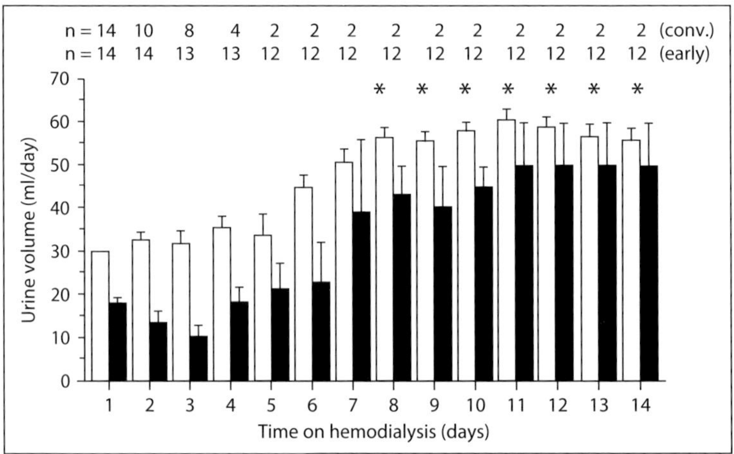

Fig. 2. Changes in urine volume after the initiation of dialysis therapy. There was no significant difference between the 2 groups. □ = Early intervention; ■ = conventional intervention. * $p < 0.05$ vs. baseline values in early intervention group. Values represent means ± SD.

Changes in Serum Creatinine (Fig. 3)

In the first 3 days after the start of continuous hemodialysis, serum creatinine increased slightly in both groups. On the eighth day and thereafter, serum creatinine decreased gradually in the early start group, and there were significant

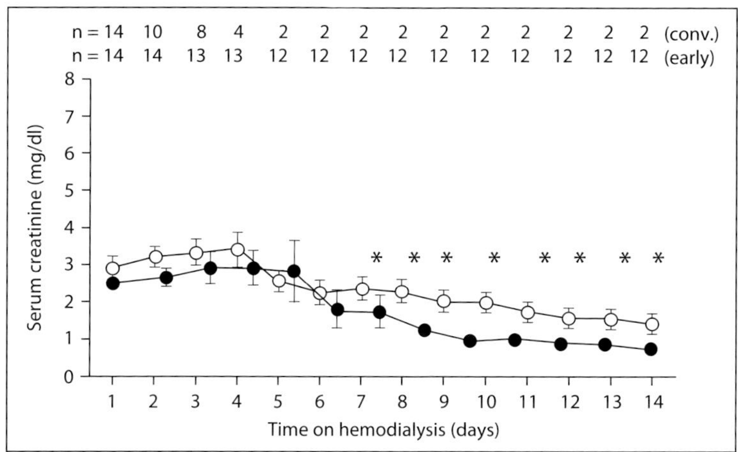

Fig. 3. Changes in serum creatinine after the initiation of dialysis therapy. There was no significant difference between 2 groups. ○ = Early intervention; ● = conventional intervention. * $p < 0.05$ vs. baseline values in early intervention group. Values represent means ± SD.

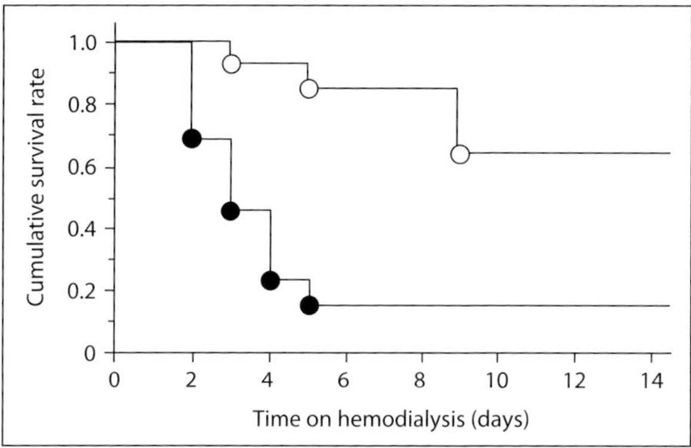

Fig. 4. Kaplan-Meier curve showing the survival rates of the 2 groups. ○ = Early intervention; ● = conventional intervention. There was a significant difference between the 2 groups ($p < 0.01$).

differences from those at the start of dialysis ($p < 0.05$). In the conventional start group, although serum creatinine showed a tendency to decrease, there was no statistical significance, likely due to the small number of subjects.

Survival Curves (Fig. 4)

The survival rate in the early start group was significantly improved relative to the conventional start group. On the 14th day, the number of survivors in the early start group was 12/14 patients. By contrast, the number of survivors in the conventional treatment group was 2/14. The 2 survivors in the conventional treatment group were weaned off dialysis on the 7th and 10th days, respectively.

In this study, continuous hemodialysis therapy was started when urinary output decreased to <30 ml/h (equivalent to ≤400 ml/day), this being the definition of oliguria. Considering that a relatively large volume of infusion is given during and after the CABG surgery, a urine output of 400 ml/day might not be enough to prevent cardiac overload. As demonstrated in the previous study of Sugahara and Suzuki [23], daily water removal of 1,500 ml or more was needed for the first 3 days after CABG surgery.

Although the timing of the start of continuous hemodialysis therapy may differ according to the individual patient's cardiac and respiratory functions, it is highly likely that decreased urinary output may be a useful index for determination of the timing of start of dialysis therapy for prevention of AKI.

Conclusion

Preoperative renal dysfunction is an important risk factor for predicting acute renal injury requiring dialysis therapy. If it occurs, AKI after cardiac surgery should be treated with the early start of continuous hemodialysis.

References

1 Schmitt H, Riehl J, Boseila A, et al: Acute renal failure following cardiac surgery: pre- and perioperative clinical features. Contrib Nephrol 1991;93:98–104.

2 Chertow GM, Levy EM, Hammermeister KE, Grover F, Daley J: Independent association between acute renal failure and mortality following cardiac surgery. Am J Med 1998;104:343–348.

3 Lassnigg A, Schmidlin D, Mouhieddine M, et al: Minimal changes of serum creatinine predict prognosis in patients after cardiothoracic surgery: a prospective cohort study. J Am Soc Nephrol 2004;15:1597–1605.

4 Rosner MH, Okusa MD: Acute kidney injury associated with cardiac surgery. Clin J Am Soc Nephrol 2006;1:19–32.

5 Kubal C, Srinivasan AK, Grayson AD, Fabri BM, Chalmers JA: Effect of risk-adjusted diabetes on mortality and morbidity after coronary artery bypass surgery. Ann Thorac Surg 2005;79:1570–1576.

6 Plume SK, O'Connor GT, Olmstead EM: As originally published in 1994: changes in patients undergoing coronary artery bypass grafting: 1987–1990. Updated in 2000. Northern New England Cardiovascular Disease Study Group. Ann Thorac Surg 2001;72:314–315.

7 Mangano CM, Diamondstone LS, Ramsay JG, Aggarwal A, Herskowitz A, Mangano DT: Renal dysfunction after myocardial revascularization: risk factors, adverse outcomes, and hospital resource utilization. The Multicenter Study of Perioperative Ischemia Research Group. Ann Intern Med 1998;128:194–203.

8 Edwards FH, Peterson ED, Coombs LP, et al.: Prediction of operative mortality after valve replacement surgery. J Am Coll Cardiol 2001;37:885–892.
9 Clark RE: The STS Cardiac Surgery National Database: an update. Ann Thorac Surg 1995;59:1376–1380, discussion 1380–1381.
10 Mehta RH, Grab JD, O'Brien SM, et al.: Bedside tool for predicting the risk of postoperative dialysis in patients undergoing cardiac surgery. Circulation 2006;114: 2208–2216, quiz 2208.
11 Abel RM, Buckley MJ, Austen WG, Barnett GO, Beck CH Jr, Fischer JE: Etiology, incidence, and prognosis of renal failure following cardiac operations: results of a prospective analysis of 500 consecutive patients. J Thorac Cardiovasc Surg 1976;71:323–333.
12 Samuels LE, Sharma S, Morris RJ, et al.: Coronary artery bypass grafting in patients with chronic renal failure: a reappraisal. J Card Surg 1996;11:128–133, discussion 134–135.
13 Chertow GM, Lazarus JM, Christiansen CL, et al: Preoperative renal risk stratification. Circulation 1997;95:878–884.
14 Weerasinghe A, Hornick P, Smith P, Taylor K, Ratnatunga C: Coronary artery bypass grafting in non-dialysis-dependent mild-to-moderate renal dysfunction. J Thorac Cardiovasc Surg 2001;121:1083–1089.
15 Anderson RJ, O'Brien M, MaWhinney S, et al: Mild renal failure is associated with adverse outcome after cardiac valve surgery. Am J Kidney Dis 2000;35:1127–1134.
16 Ryckwaert F, Boccara G, Frappier JM, Colson PH: Incidence, risk factors, and prognosis of a moderate increase in plasma creatinine early after cardiac surgery. Crit Care Med 2002;30:1495–1498.
17 Wang F, Dupuis JY, Nathan H, Williams K: An analysis of the association between preoperative renal dysfunction and outcome in cardiac surgery: estimated creatinine clearance or plasma creatinine level as measures of renal function. Chest 2003;124:1852–1862.
18 Urzua J, Troncoso S, Bugedo G, et al.: Renal function and cardiopulmonary bypass: effect of perfusion pressure. J Cardiothorac Vasc Anesth 1992;6:299–303.
19 Wijeysundera DN, Karkouti K, Dupuis JY, et al: Derivation and validation of a simplified predictive index for renal replacement therapy after cardiac surgery. JAMA 2007;297:1801–1809.
20 Thakar CV, Arrigain S, Worley S, Yared JP, Paganini EP: A clinical score to predict acute renal failure after cardiac surgery. J Am Soc Nephrol 2005;16:162–168.
21 Detsky AS, Abrams HB, McLaughlin JR, et al.: Predicting cardiac complications in patients undergoing non-cardiac surgery. J Gen Intern Med 1986;1:211–219.
22 Goldman L, Caldera DL, Nussbaum SR, et al: Multifactorial index of cardiac risk in non-cardiac surgical procedures. N Engl J Med 1977;297:845–850.
23 Sugahara S, Suzuki H: Early start on continuous hemodialysis therapy improves survival rate in patients with acute renal failure following coronary bypass surgery. Hemodial Int 2004;8:320–325.

Hiromichi Suzuki, PhD, MD
Department of Nephrology, Saitama Medical University
Morohongo 38, Moroyamamachi Irumagun
Saitama 350-0495 (Japan)
Tel. +81 492 76 1620, Fax +81 492 76 1620, E-Mail iromichi@saitama-med.ac.jp

Suzuki H, Hirasawa H (eds): Acute Blood Purification.
Contrib Nephrol. Basel, Karger, 2010, vol 166, pp 40–46

Septic Acute Renal Failure

Tsuyoshi Mori · Tomoharu Shimizu · Tohru Tani

Department of Surgery, Shiga University of Medical Science, Ōtsu, Japan

Abstract

Acute renal failure (ARF) is the rapid loss of the renal filtration function, which is characterized by metabolic acidosis, high potassium levels, a body fluid imbalance, and so on. The overall mortality rate of ARF is about 45%; however, the mortality rate of sepsis-induced ARF is about 70%. In addition, sepsis is the most common trigger of ARF. Little is known about the pathogenesis of septic ARF, although renal hypoperfusion and ischemia have been proposed as being central. Blood purification therapies for septic ARF include the elimination of pathogenesis, such as endotoxin or mediators that contribute to ARF, and renal replacement therapy (RRT). The adsorption of endotoxin with direct hemoperfusion using polymyxin-B immobilized fiber makes the urinary output increase, while also improving renal function. It would seem logical to initiate RRT earlier rather than later, especially in rapidly developing symptomatic oliguric renal failure with metabolic derangement. Continuous RRT (CRRT) has an advantage over intermittent RRT in that it provides greater hemodynamic stability, easier fluid removal and greater flexibility in providing parenteral nutrition as a result of a greater control over the fluid balance. CRRT may be able to reduce chronic dialysis dependence. Patients with sepsis and ARF are hypercatabolic. Some studies have suggested that increased doses of dialysis improve survival in patients who are hypercatabolic and have ARF. The increase in the ultrafiltration rate may, however, be associated with some difficulties, namely cost and labor. The mechanisms of septic ARF therefore need to be further elucidated and the potential of RRT in improving the mortality associated with ARF needs to be established.

Acute renal failure (ARF) is the rapid loss of the renal filtration function, which is characterized by metabolic acidosis, high potassium levels, a body fluid imbalance, and so on. This condition is usually marked by a rise in the serum creatinine concentration or blood urea nitrogen (BUN) concentration. However, no common clear criteria regarding ARF had been established, and thus a consensus definition was required. The Acute Dialysis Quality Initiative therefore developed a consensus definition of acute kidney injury (AKI) that goes under the acronym of RIFLE (risk, injury, failure, loss, end-stage renal failure). The term AKI is considered to represent the entire spectrum of ARF. The diagnostic criteria for AKI are based on acute alterations in serum creatinine or urine output. The diagnostic criteria for AKI are: (1) an abrupt (within 48 h) reduction in kidney function currently defined as an absolute increase in serum creatinine of ≥0.3 mg/dl (≥26.4 μmol/l), (2) an increase in serum creatinine of ≥50% (1.5-fold from baseline), or (3) a reduction in urine output (documented oliguria of <0.5 ml/kg/h for more than 6 h) [1].

Sepsis is the most common trigger of ARF. ARF occurs in approximately 19% of patients with moderate sepsis, 23% with severe sepsis, and 51% with septic shock when blood cultures are positive [2]. In addition, septic shock is the most common contributing factor to ARF. The frequency of its contribution to the development of ARF is around 50% [3]. The overall mortality rate of ARF is about 45%; however, the mortality rate of sepsis-induced ARF is about 70% [4]. Despite our increasing ability to support vital organs and resuscitate patients, the incident and mortality rates of septic ARF remain high. It is therefore very important to elucidate septic ARF and to implement rational treatments for patients with this disease.

Pathophysiology

A major paradigm that has been derived from observations in animals and humans with hypodynamic shock (hemorrhagic, cardiogenic or even septic) is that AKI is due to renal ischemia. This construct therefore implies that the restoration of adequate renal blood flow (RBF) should be the primary means of renal protection in critically ill patients. Acute tubular necrosis has been histopathologically confirmed in shock-associated ARF. Systemic vasodilatation and enhancement of vascular permeability is induced by nitric oxide, bradykinin, various cytokines and endogenous marijuana – especially in septic shock. As a result, the systemic blood flow and vascular resistance decrease, as does renal blood flow. Vasopressin is secreted to maintain systemic blood flow, thus sympathetic nervous system activity and renin-angiotensin-aldosterone system activity are enhanced, and endothelin is activated; however, RBF decreases because of the imbalance in the distribution of blood flow and renal vasoconstriction. On the other hand, hypercoagulability occurs with inflammation and microthrombi develop in the glomerulus, encouraging further renal ischemia. In addition, the vascular endothelium becomes injured, and, as a result, it prevents the synthesis of nitric oxide which distends blood vessels in the kidney. ARF results from these overlapping phenomena.

Bellomo et al. [5] continuously measured cardiac output and RBF in sheep while a high cardiac output septic state was induced by the infusion of *Escherichia coli*. They were able to show that the RBF markedly increased, while the renal vascular resistance markedly decreased in hyperdynamic sepsis in a conscious large mammal. In this setting, the glomerular filtration rate was markedly diminished, with a 3-fold increase in serum creatinine concentration and an equivalent decrease in creatinine clearance. These observations suggest that changes in renal vascular activity (vasodilation) may be important in the loss of glomerular filtration pressure during the first 24–48 h of sepsis. They also showed the expression of early phase proapoptotic proteins, such as BAX, as well as counterbalancing antiapoptotic proteins, such as Bcl-xL in the tubular cells after only 3 h of sepsis induced by intravenous injection of *E. coli*; thus, indicating an early activation of the apoptotic cascade in the septic kidney. This report implies that acute tubular apoptosis is involved in AKI.

Therapy

Studies have shown that there are hundreds of mediators involved in sepsis. Microbial components – such as endotoxin, peptidoglycan, peptidoglycan-associated lipoprotein, lipoteichoic acid, β-glucans and other membrane proteins – have been reported to activate inflammatory cascades during microbial infections. Blood purification therapies in septic ARF include the elimination of pathogens, such as endotoxin or mediators that contribute to ARF, and renal replacement therapy (RRT). The adsorption of endotoxin with direct hemoperfusion using polymyxin-B immobilized fiber (PMX-DHP) makes urinary output increase while also improving renal function. The first clinical report on PMX treatment was published in 1994 by Aoki et al. [6] who reported that 16 patients were safely and successfully treated with a polymyxin B immobilized fiber column (Toraymyxin®). Toraymyxin was developed in Japan for the purpose of selectively adsorbing endotoxin, with these columns having being used clinically in the treatment of sepsis in over 30,000 patients under the Japanese National Health Insurance system. Vincent et al. [7] demonstrated that the need for continuous RRT (CRRT) after study entry decreased in the PMX group.

Nakamura et al. [8] reported that the increased urinary NAG/creatinine ratio in patients with severe sepsis may reflect proximal tubular dysfunction. PMX is effective in reducing proximal tubular dysfunction, in part owing to reduced plasma endotoxin levels.

Cantaluppi et al. [9] also reported that SOFA (sequential organ failure assessment score) and RIFLE scores, proteinuria and urine tubular enzymes were all significantly reduced after PMX treatment. Loss of plasma-induced polarity and permeability of cell cultures was abrogated with the plasma of patients treated with PMX. These results were associated with a preserved expression of molecules crucial for tubular and glomerular functional integrity. These reports show that the attenuation of 'the upstream' improves septic ARF. This evidence is very useful for understanding how septic ARF progresses. On the other hand, there are some blood purification therapies to eliminate cytokines which are produced in endotoxemia, e.g. CHDF using polymethylmethacrylate membrane [10] or polycrylonitrile membrane and adsorption using CTR-001 [11, 12]. These topics are covered in greater detail in other chapters of this book.

Next we will discuss RRT in detail. The key issues in the use of RRT in septic ARF are (1) the timing of initiation, (2) continuous (CRRT) or intermittent (IRRT) RRT, and (3) dialysis dose. The specific indications for the initiation of RRT in septic ARF remain controversial.

Timing of the Initiation

A prospective cohort study showed that the risk of death was lower when dialysis was started at lower blood urea levels [13]. Palevsky [14] reviewed a series of retrospective case series and observational studies conducted from the 1950s through the early 1970s that compared 'early' initiation of hemodialysis (defined by BUN concentrations ranging from <93 mg/dl to levels of approximately 150 mg/dl) to 'late' initiation of therapy (defined by BUN levels of 163 to >200 mg/dl) and these studies all demonstrated improved survival with the earlier initiation of hemodialysis. Although the initiation of RRT in patients with progressive azotemia prior to the development of overt uremia is effective, the evidence regarding the ideal timing of dialysis is lacking. Three different studies have shown inconclusive results [15–17]. The specific creatinine or BUN level at which RRT should be commenced in ARF is difficult to define. Biomarkers more sensitive than the rise in serum creatinine or BUN concentration associated with ARF will be necessary to achieve early intervention. It would therefore seem logical to initiate RRT earlier rather than later, especially in rapidly developing symptomatic oliguric renal failure with metabolic derangement.

Continuous or Intermittent RRT?

CRRT has an advantage over IRRT in that it provides greater hemodynamic stability, easier fluid removal and greater flexibility in providing parenteral nutrition as a result of a greater control over the fluid balance [18], which might prolong the course of ARF and thereby increase mortality. Bagshaw et al. [19], however, reported that there was no significant difference in outcome (in terms of mortality or renal recovery) between CRRT and IRRT. Pannu et al. [20] also showed that the current state of evidence does not indicate any clear advantage in the use of CRRT over IRRT in septic ARF. Uchino [21] reported that randomized controlled trials conducted so far do not support the effectiveness of CRRT over IRRT in relation to renal recovery; on the other hand, CRRT may be able to reduce chronic dialysis dependence.

Dialysis Dose

Patients with sepsis and ARF are hypercatabolic. Some studies suggest that increased doses of dialysis improve survival in patients who are hypercatabolic and have ARF. Hemofiltration produces better survival rates than peritoneal dialysis in patients with ARF associated with malaria and other infections [22]. CHF and CHDF were introduced as CRRT to critically ill patients at first. They were expected to remove cytokines and a variety of mediators that cause of multiple organ failure. Ronco et al. [23] performed a prospective randomized trial on 425 patients with ARF. They showed that the ultrafiltration rate of 35 or 45 ml/kg/h in comparison to 20 ml/kg/h improves survival in ARF using continuous venovenous hemofiltration. Better survival was observed with an ultrafiltration rate of 45 ml ml/kg/h than with a rate of 35 ml/kg/h in patients with sepsis-related ARF [24]. They proposed the concept of the peak concentration hypothesis. This suggests that continuous renal replacement therapies, due to their continuity and unspecific capacity of removal, might be beneficial in cutting the peaks of the concentrations of both pro- and anti-inflammatory mediators, restoring a situation of immunohomeostasis [25]. The increase in the ultrafiltration rate may, however, be associated with some difficulties, namely cost and labor. To resolve these difficulties, it is necessary to convert it to online hemodiafiltration or devise a large pore-size hemofilter to improve the extraction coefficient of large molecules because the molecular weight of cytokines lies between 8 and 30 kDa. A prospective randomized multicenter trial of high volume CHF is now in progress in Europe (IVOIRE, National Clinical Trail No. 00241228). This trial compares 2 treatments in patients with septic shock complicated with ARF admitted to an ICU. One group will be treated by early high volume hemofiltration (70 ml/kg/h) and the second group by standard volume hemofiltration (35 ml/kg/h). The main outcome will be 1-month mortality. This study will be complete in April 2010.

The development of hemodialysis has led to a dramatic decrease in mortality of patients with ARF clinical syndrome from about 90% to about 50%. However, since these initial successes, the mortality level has remained high for the last several decades. The mechanisms of septic ARF therefore need to be further elucidated and the potential of RRT in improving the mortality associated with ARF needs to be established.

References

1 Bagga A, Bakkaloglu A, Devarajan P, Mehta RL, Kellum JA, Shah SV, Molitoris BA, Ronco C, Warnock DG, Joannidis M, Levin A: Improving outcomes from acute kidney injury: report of an initiative. Pediatr Nephrol 2007;22:1655–1658.

2 Rangel-Frausto MS, Pittet D, Costigan M, Hwang T, Davis CS, Wenzel RP: The natural history of the systemic inflammatory response syndrome (SIRS): a prospective study. JAMA 1995;273:117–123.

3 Uchino S, Kellum JA: Acute renal failure in critically ill patients. JAMA 2005;294:813–818.

4 Schrier RW, Wang W: Acute renal failure and sepsis. N Engl J Med 2004;351:159–169.

5 Bellomo R, Wan L, Langenberg C, May C: Septic acute kidney injury: new concepts. Nephron Exp Nephrol 2008;109:e95–e100.

6 Aoki H, Kodama M, Tani T, Hanasawa K: Treatment of sepsis by extracorporeal elimination of endotoxin using polymyxin B-immobilized fiber. Am J Surg 1994;167: 412–417.

7 Vincent JL, Laterre PF, Cohen J, Burchardi H, Bruining H, Lerma FA, Wittebole X, De Backer D, Brett S, Marzo D, Nakamura H, John S: A pilot-controlled study of a polymyxin B-immobilized hemoperfusion cartridge in patients with severe sepsis secondary to intra-abdominal infection. Shock 2005;23:400–405.

8 Nakamura T, Kawagoe Y, Matsuda T, Ueda Y, Koide H: Effects of polymyxin B immobilized fiber on urinary N-acetyl-beta-glucosaminidase in patients with severe sepsis. ASAIO J 2004;50:563–567.

9 Cantaluppi V, Assenzio B, Pasero D, Romanazzi GM, Pacitti A, Lanfranco G, Puntorieri V, Martin EL, Mascia L, Monti G, Casella G, Segoloni GP, Camussi G, Ranieri VM: Polymyxin-B hemoperfusion inactivates circulating proapoptotic factors. Intensive Care Med 2008;34:1638–1645.

10 Matsuda K, Moriguchi T, Harii N: Comparison of efficacy between continuous hemodiafiltration with a PMMA membrane hemofilter and a PAN membrane hemofilter in the treatment of a patient with septic acute renal failure. Transfus Apher Sci 2009;40:49–53.

11 Taniguchi T, Hirai F, Takemoto Y, Tsuda K, Yamamoto K, Inaba H, Sakurai H, Furuyoshi S, Tani N: A novel adsorbent of circulating bacterial toxins and cytokines: the effect of direct hemoperfusion with CTR column for the treatment of experimental endotoxemia. Crit Care Med 2006;34:800–806.

12 Taniguchi T, Kurita A, Yamamoto K, Inaba H: Comparison of a cytokine adsorbing column and an endotoxin absorbing column for the treatment of experimental endotoxemia. Transfus Apher Sci 2009;40:55–59.

13 Liu KD, Himmelfarb J, Paganini E, Ikizler TA, Soroko SH, Mehta RL, Chertow GM: Timing of initiation of dialysis in critically ill patients with acute kidney injury. Clin J Am Soc Nephrol 2006;1:915–919.

14 Palevsky PM: Timing and dose of continuous renal replacement therapy in acute kidney injury. Crit Care 2007;11:232.

15 Abichandani R, Pereira BJ: Effects of different doses in continuous veno-venous haemofiltration on outcomes of acute renal failure: a prospective randomised trial, by Ronco C, Bellomo R, Homel P, Brendolan A, Dan M, Piccinni P, La Greca G. Lancet 355:26–30, 2000. Semin Dial 2001;14:233–234.

16 Sugahara S, Suzuki H: Early start on continuous hemodialysis therapy improves survival rate in patients with acute renal failure following coronary bypass surgery. Hemodial Int 2004;8:320–325.

17 Bouman CS, Oudemans-Van Straaten HM, Tijssen JG, Zandstra DF, Kesecioglu J: Effects of early high-volume continuous venovenous hemofiltration on survival and recovery of renal function in intensive care patients with acute renal failure: a prospective, randomized trial. Crit Care Med 2002; 30:2205–2211.

18 Rajapakse S, Rodrigo C, Rajapakse A, Kirthinanda D, Wijeratne S: Renal replacement therapy in sepsis-induced acute renal failure. Saudi J Kidney Dis Transpl 2009;20: 553–559.

19 Bagshaw SM, Berthiaume LR, Delaney A, Bellomo R: Continuous versus intermittent renal replacement therapy for critically ill patients with acute kidney injury: a meta-analysis. Crit Care Med 2008;36:610–617.

20 Pannu N, Klarenbach S, Wiebe N, Tonelli M, Alberta Kidney Disease Network: Renal replacement therapy in patients with acute renal failure: a systematic review. JAMA 2008;299:793–805.

21 Uchino S: Choice of therapy and renal recovery. Crit Care Med 2008;36:S238–S242.

22 Phu NH, Hien TT, Mai N, Chau TT, Chuong LV, Loc PP, Winearls C, Farrar J, White N, Day N: Hemofiltration and peritoneal dialysis in infection-associated acute renal failure in Vietnam. N Engl J Med 2002;347:895–902.

23 Ronco C, Bellomo R, Homel P, Brendolan A, Dan M, Piccinni P, La Greca G: Effects of different doses in continuous venovenous haemofiltration on outcomes of acute renal failure: a prospective randomized trial. Lancet 2000;356:26–30.

24 Tonelli M, Manns B, Feller-Kopman D: Acute renal failure in the intensive care unit: a systematic review of the impact of dialytic modality on mortality and renal recovery. Am J Kidney Dis 2002;40:875–885.

25 Ronco C, Bonello M, Bordoni V, Ricci Z, D'Intini V, Bellomo R, Levin NW: Extracorporeal therapies in non-renal disease: treatment of sepsis and the peak concentration hypothesis. Blood Purif. 2004;22: 164–174.

Tsuyoshi Mori, MD, PhD
Department of Surgery, Shiga University of Medical Science,
Seta, Tsukinowa-cho
Otsu, Shiga, 520–2192 (Japan)
Tel. +81 77 548 2238, Fax +81 77 548 2240, E-Mail t252m@belle.shiga-med.ac.jp

Suzuki H, Hirasawa H (eds): Acute Blood Purification.
Contrib Nephrol. Basel, Karger, 2010, vol 166, pp 47–53

Non-Renal Indications for Continuous Renal Replacement Therapy: Current Status in Japan

Shigeto Oda · Tomohito Sadahiro · Yo Hirayama · Masataka Nakamura · Eizo Watanabe · Yoshihisa Tateishi · Hiroyuki Hirasawa

Department of Emergency and Critical Care Medicine, Graduate School of Medicine, Chiba University, Chiba, Japan

Abstract

Continuous renal replacement therapy (CRRT) has been extensively used in Japan as renal support for critically ill patients managed in the ICU. In Japan, active research has also been conducted on non-renal indications for CRRT, i.e. the use of CRRT for purposes other than renal support. Various methods of blood purification have been attempted to remove inflammatory mediators, such as cytokines, in patients with severe sepsis or septic shock. In these attempts, efficacy was demonstrated for continuous hemodiafiltration (CHDF) using a polymethyl methacrylate (PMMA) membrane hemofilter which is capable of adsorbing and removing various cytokines, plasma diafiltration, and online CHDF. Furthermore, a recently developed cytokine-adsorbing column is now under clinical evaluation. Definite evidence for the efficacy of CRRT for non-renal indications has not been established. In evaluating the efficacy of CRRT for non-renal indications, it is essential to focus on patients subjected to be studied, such as severe sepsis or septic shock, and to evaluate its indication, commencement, termination of therapy and also its therapeutic effects based on analysis of blood levels of the target substances to be removed (e.g. cytokines). IL-6 blood level appears to be useful as a variable for this evaluation. It is expected that evidence endorsing the validity of these methods now being attempted in Japan will be reported near future.

Continuous arteriovenous hemofiltration, first reported in 1977 by Kramer et al. [1], was clinically introduced to Japan in the middle of the 1980s as a method of continuous renal replacement therapy (CRRT) for the use in the ICU. Initially, it was performed as veno-venous continuous hemofiltration. It was subsequently modified to continuous hemodiafiltration (CHDF) to improve its capacity for removal of solutes. At present, CHDF is extensively used in Japan as renal support for critically ill patients managed in the ICU [2, 3]. There are several factors for widespread use of CRRT in Japan. One is the development of a console designed specifically for the use of CRRT at the bedside in the ICU and the other is introduction of nafamostat mesilate as an anticoagulant for extracorporeal circulation. Nafamostat mesilate is a synthetic serine protease inhibitor developed in Japan and which possesses a selective anticoagulant property and short half life. Since introduction of nafamostat mesilate, hemorrhagic complications due to prolonged use of the anticoagulant dramatically decreased and CRRT could be performed safely even in patients showing bleeding tendency [4].

In 1993, Bellomo et al. [5] reported removal of various cytokines with CRRT. Since then, CRRT has been used for so-called 'non-renal indications' to remove humoral mediators such as inflammatory cytokines, and active research on non-renal indications for CRRT has been carried out in Japan. At present, non-renal indications for CRRT approved by National Health Insurance in Japan include fulminant hepatic failure as an artificial liver support and severe acute pancreatitis even in patients without renal failure. More recently, various methods of blood purification aimed at removing inflammatory mediators have been attempted for treatment of acute respiratory distress syndrome, severe sepsis and septic shock.

This article will review the current status of non-renal indications for CRRT in Japan.

Blood Purification Advances in Japan

Blood purifications have advanced primarily in the form of hemodialysis for patients with chronic renal failure. At present, there are more than 270,000 patients on chronic hemodialysis in Japan, including those receiving hemodialysis, hemodiafiltration, and other related treatments. The results of these treatments for chronic renal failure have been excellent, and the Japanese results are probably the best in the world [6]. Furthermore, thanks to advances in medical engineering technology, various techniques of blood purification have been developed and applied clinically to patients with a wide variety of diseases. Plasma exchange is used not only for the treatment of fulminant hepatic failure but also for thrombotic thrombocytopenic purpura and toxic epidermal necrolysis. In addition, as methods for more selective removal of pathogenic substances, double-filtration plasmapheresis (DFPP) and plasma adsorption have been developed and applied to the treatment of autoimmune diseases, neuromuscular diseases, hyperlipidemia, and other conditions. Recently, DFPP has been approved by National Health Insurance as a measure of the removal of hepatitis C virus for the treatment of hepatitis C patients refractory to antiviral treatment.

Table 1. Acute blood purification in critical care currently performed in Japan

Blood purification technique	Indication
Continuous hemofiltration/ continuous hemodiafiltration	Acute kidney injury Severe sepsis/septic shock Severe acute pancreatitis Fulminant hepatic failure
Plasma exchange	Fulminant hepatic failure Thrombotic thrombocytopenic purpura Toxic epidermal necrolysis
Direct hemoperfusion	Drug intoxication (charcoal DHP) Endotoxic shock (PMX-DHP)

Furthermore, a column capable of adsorbing and removing β2-microglobulin (a substance responsible for dialysis related amyloidosis) by direct hemoperfusion (DHP) and an endotoxin-adsorbing column (PMX-DHP) have also been developed and introduced clinically. Additionally, a column capable of removing leukocytes in circulating blood (leukocytapheresis and granulocytapheresis) has been developed and used for the treatment of inflammatory bowel diseases, malignant rheumatoid arthritis and other conditions.

In the field of critical care, CRRT by means of continuous hemofiltration, CHDF, PMX-DHP and other techniques have been attempted in patients under various pathophysiologic conditions to remove humoral mediators (particularly cytokines) continuously produced in the body. Under the current health insurance system in Japan, fulminant hepatic failure and severe acute pancreatitis are covered as non-renal indications for CRRT. Furthermore, attempts are actively being made to apply CRRT clinically in patients with severe sepsis and septic shock (table 1).

In Western countries, high-volume hemofiltration [7] and high-flux hemofiltration [8] have been attempted to remove inflammatory mediators, since the study by Ronco et al. [9] demonstrated significant improvement in the outcome of patients with acute renal failure in response to higher dose of blood purification.

In Japan, attempts to remove inflammatory mediators have been conducted employing: (1) CHDF using polymethyl methacrylate membrane hemofilter designed to remove cytokines primarily through adsorption to the hemofilter membrane (PMMA-CHDF); (2) plasma diafiltration, hemodiafiltration using large-pore plasma fractionators (originally designed for the second filter of DFPP); (3) high-volume, high-flow dialysate hemodiafiltration, and (4) online CHDF. The other method besides CRRT is DHP using a cytokine adsorbing column (table 2). These methods are briefly introduced in the following sections.

Table 2. Blood purification techniques for cytokine removal in non-renal indications

PMMA-CHDF
Plasma diafiltration
High-flow-volume CHDF
Online CHDF
Cytokine adsorbing column

PMMA-CHDF

A variety of membrane materials are available for the hemofilter used in CRRT. Among them, PMMA membrane has been shown to be capable of nonspecifically adsorbing low-molecular-weight proteins [10]. Hemodiafiltration with a PMMA membrane hemofilter (PMMA-CHDF) can effectively and continuously remove various cytokines from circulating blood, and its clinical efficacy has been reported in patients with various diseases [11, 12]. It has been reported that PMMA-CHDF reduced blood IL-6 levels significantly in patients with septic shock, resulting in improvement of hemodynamics and outcome [13]. More recently, attempts at high-flow-volume PMMA-CHDF (high-volume filtration and high-flow-rate dialysis with a large membrane area PMMA dialyzer to increase the cytokine-removing capacity of PMMA) and double PMMA-CHDF (simultaneous application of 2 systems of PMMA-CHDF) have been made in patients with refractory septic shock, and clinical efficacy has been reported. Recently, efficacy of cytokine-absorbing hemofiltration using AN-69 (which adsorbs and removes low-molecular-weight proteins, comparable to that of the PMMA membrane hemofilter) has been reported [14]. Because removal by adsorption onto hemofilter membrane does not require large volume of filtration or dialysis, it can be applied even to patients with limited blood flow and is advantageous in terms of cost. It would thus be a promising new method.

Plasma Diafiltration

Plasma diafiltration involves filtration and dialysis using a membrane pore size 0.008–0.01 μm in the Evacure EC-2A plasma separator, which is made of ethylene vinylalcohol membrane. This device was originally designed for use as the secondary filter (i.e. plasma fractionator) for DFPP. The sieving coefficient for albumin of this filter is 0.3. Therefore, low-molecular-weight proteins (those with molecular weights lower than that of albumin) and a portion of albumin-bound substances can be removed with this membrane primarily by filtration.

However, to perform plasma diafiltration, it is necessary to administer fresh frozen plasma or albumin as replacement fluid. At present, clinical studies on plasma diafiltration are under way to evaluate its usefulness as an artificial liver support for patients with hepatic failure [15] and as a means of removing cytokines in patients with septic shock.

Online CHDF

When high-volume hemofiltration is performed with CRRT, filtration volume is limited by the blood flow rate as well as massive use of bicarbonate endotoxin-free replacement fluid. With online CHDF, however, the volume of filtration can be increased without limit, since this technique uses a portion of the endotoxin-free dialysate as replacement fluid. It has been reported that online CHDF could improve the outcome of patients with septic shock [16]. One shortcoming of online CHDF is that it cannot be performed in ordinary ICUs because it requires an endotoxin-free dialysate supply system that consists of a water purifying facility and endotoxin-removing filter, and the purity of dialysate must be routinely monitored.

Cytokine-Adsorbing Column

A column selectively adsorbing cytokines through DHP (CTR-001) has been developed and is now under clinical evaluation. This column was obtained by modification of the adsorptive column Lixelle® previously developed for adsorption of β2-microglobulin, which is a causative substance for dialysis-related amyloidosis [17]. It has been shown that this column can adsorb and remove various cytokines [18]. At present, a phase III study of this column in septic shock is under way, and clinical evaluation will appear in the near future.

Future Perspectives

The results of recently reported multicenter randomized controlled trials (RCTs) on CRRT enhancing blood purification dose [19, 20] have been disappointing. Survival rates were not improved despite increases in intensity of CRRT, unlike the results seen in the Ronco et al. study [9]. However, all of these RCTs involved patients with acute kidney injury, and none was designed to evaluate efficacy of removal of the inflammatory mediators (cytokines and related substances) that play important roles in severe sepsis and septic shock. Organ dysfunction such as acute kidney injury develops as a result of severe and persistent hypercytokinemia, and the removal of cytokines after development of organ dysfunction would not be expected to be effective. In evaluating efficacy of removal of cytokines, it is necessary to confine study subjects to patients with severe sepsis or septic shock. It is ideal in such a study to enroll only patients with demonstrably high blood levels of cytokines. Furthermore, in examining the clinical efficacy of such treatment, it is advisable to set the goal at alleviation of the target condition (hypercytokinemia) and the resultant improvement of organ perfusion or organ dysfunction.

In Japan, several institutions employ a rapid assay system of IL-6 blood level in the clinical laboratory and use it to judge non-renal indications for CRRT, to evaluate treatment effect, and to decide the timing of weaning from treatment on an individual basis. Among various cytokines, IL-6 is one that enables reliable measurement, and it has been reported that IL-6 blood level well reflects the intensity of hypercytokinemia and the severity of the patient's condition [21]. If measurement of these biomarkers is performed routinely at many facilities, determination of the optimal timing of CRRT as a non-renal indication, objective evaluation of the efficacy of treatment would be possible.

Recently, the efficacy of PMX-DHP for septic shock caused by abdominal sepsis has been demonstrated in a multicenter RCT [22]. However, there is as yet no reported evidence for the efficacy of CRRT in non-renal indications. As mentioned above, various blood purifications for non-renal indications are widely performed in Japan and their efficacy is reported. However, most reports are published only in Japanese. It is desirable that the efficacy of various methods of blood purification applied ibn non-renal indications in Japan would be more frequently published in English and be endorsed by RCTs in the near future.

References

1 Kramer P, Wigger W, Rieger J, Matthei D, Scheler F: Arteriovenous hemofiltration: a new and simple method for treatment of over-hydrated patients resistant to diuretics. Klin Wochenschr 1977;55:1121–1122.

2 Hirasawa H, Sugai T, Ohtake Y, Oda S, Matsuda K, Kitamura N: Blood purification for prevention and treatment of multiple organ failure. World J Surg 1996;20:482–486.

3 Oda S, Hirasawa H, Shiga H, Nakanishi K, Matsuda K, Nakamura M: Continuous hemofiltration/hemodiafiltration in critical care. Ther Apher 2002;6:193–198.

4 Ohtake Y, Hirasawa H, Sugai T, Oda S, Shiga H, Matsuda K, Kitamura N: Nafamostat mesylate as anticoagulant in continuous hemofiltration and continuous hemodiafiltration. Contrib Nephrol 1991;93:215–217.

5 Bellomo R, Tipping P, Boyce N: Continuous veno-venous hemofiltration with dialysis removes cytokines from the circulation of septic shock. Crit Care Med 1993;21:522–526.

6 Nakai S, Masakane I, Shigematsu T, Hamano T, Yamagata K, Watanabe Y, Itami N, Ogata S, Kimata N, Shinoda T, Syouji T, Suzuki K, Taniguchi M, Tsuchida K, Nakamoto H, Nishi S, Nishi H, Hashimoto S, Hasegawa T, Hanafusa N, Fujii N, Marubayashi S, Morita O, Wakai K, Wada A, Iseki K, Tsubakihara Y: An Overview of Regular Dialysis Treatment in Japan (As of 31 December 2007). Ther Apher Dial 2009;13:457–504.

7 Honore PM, Jamez J, Wauthier M, Lee PA, Dugernier T, Pirenne B, Hanique G, Matson JR : Prospective evaluation of short-term, high-volume isovolemic hemofiltration on the hemodynamic course and outcome in patients with intractable circulatory failure resulting from septic shock. Crit Care Med 2000;28:3581–3587.
8 Honore PM, Joannes-Boyau O, Gressens B: Blood and plasma treatments: the rationale of high-volume hemofiltration; in Ronco C, Bellomo R, Kellum JA (eds): Acute Kidney Injury. Contrib Nephrol, Basel, Karger, 2007, vol 156, pp 387–395.
9 Ronco C, Bellomo R, Homel P, Brendolan A, Dan M, Piccinni P, La Greca G : Effects of different doses in continuous veno-venous haemofiltration on outcomes of acute renal failure: a prospective randomised trial. Lancet 2000;355:26–30.
10 Yamashita AC, Tomisawa N: Importance of membrane materials for blood purification devices in critical care. Transfus Apher Sci 2009;40:23–31.
11 Matsuda K, Hirasawa H, Oda S, Shiga H, Nakanishi K: Current topics on cytokine removal technologies. Ther Apher 2001;5: 306–314.
12 Nakada T, Hirasawa H, Oda S, Shiga H, Matsuda K: Blood purification for hypercytokinemia. Transfus Apher Sci 2006;35:253–264.
13 Nakada T, Oda S, Matsuda K, Sadahiro T, Nakamura M, Abe R, Hirasawa H: Continuous hemodiafiltration with PMMA hemofilter in the treatment of patients with septic shock. Mol Med 2008;14:257–263.
14 Haase M, Silvester W, Uchino S, Goldsmith D, Davenport P, Tipping P, Boyce N, Bellomo R: A pilot study of high-adsorption hemofiltration in human septic shock. Int J Artif Organs 2007;30:108–117.
15 Nakae H, Igarashi T, Tajimi K, Noguchi A, Takahashi I, Tsuchida S, Takahashi T, Asanuma Y: A case report of pediatric fulminant hepatitis treated with plasma diafiltration. Ther Apher Dial 2008;12:329–332.
16 Kawanishi H: On-line continuous hemodiafiltration in sepsis. Trans Aher Sci 2006;35: 265–269.
17 Furuyoshi S, Nakatani M, Taman J, Kutsuki H, Takata S, Tani N: New adsorption column (Lixelle) to eliminate beta2-microglobulin for direct hemoperfusion. Ther Apher 1998; 2:13–17.
18 Taniguchi T, Kurita A, Yamamoto K, Inaba H: Comparison of a cytokine adsorbing column and an endotoxin absorbing column for the treatment of experimental endotoxemia. Transfus Apher Sci 2009;40:55–59.
19 The VA/NIH Acute Renal Failure Trial Network: Intensity of renal support in critically ill patients with acute kidney injury. N Engl J Med 2008;359:7–20.
20 The RENAL Replacement Therapy Study Investigators: Intensity of continuous renal-replacement therapy in critically ill patients. N Engl J Med 2009;361:1627–1638.
21 Oda S, Hirasawa H, Shiga H, Nakanishi K, Matsuda K, Nakamua M: Sequential measurement of IL-6 blood levels in patients with systemic inflammatory response syndrome (SIRS)/sepsis. Cytokine 2005;29:169–175.
22 Cruz DN, Antonelli M, Fumagalli R, Foltran F, Brienza N, Donati A, Malcangi V, Petrini F, Volta G, Bobbio Pallavicini FM, Rottoli F, Giunta F, Ronco C: Early use of polymyxin B hemoperfusion in abdominal septic shock: the EUPHAS randomized controlled trial. JAMA 2009;301:2445–2452.

Shigeto Oda
1-8-1 Inohana, Chuou-ku, Chiba City
Chiba 260-8677 (Japan)
Tel. +81 043 226 2341, Fax +81 043 226 2371, E-Mail odas@faculty.chiba-u.jp

Suzuki H, Hirasawa H (eds): Acute Blood Purification.
Contrib Nephrol. Basel, Karger, 2010, vol 166, pp 54–63

Continuous Hemodiafiltration Using a Polymethyl Methacrylate Membrane Hemofilter for Severe Acute Pancreatitis

Ryuzo Abe · Shigeto Oda · Koichiro Shinozaki · Hiroyuki Hirasawa

Department of Emergency and Critical Care Medicine, Chiba University Graduate School of Medicine, Chiba, Japan

Abstract

It has been reported that hypercytokinemia plays a pivotal role in the pathophysiology of severe acute pancreatitis (SAP). In our previous reports, continuous hemodiafiltration (CHDF) using a polymethyl methacrylate (PMMA) membrane hemofilter (PMMA-CHDF) was found to be capable of efficiently removing various cytokines from circulating blood. The present study was undertaken to evaluate the efficacy of PMMA-CHDF aimed at cytokine removal in the treatment of SAP. Patients with blood IL-6 level ≥400 pg/ml were considered indicated for initiation of PMMA-CHDF based on our previous data. Among the patients enrolled in the present study, there were significant differences in APACHE II sore, JMHLW (Japanese Ministry of Health, Labour, and Welfare) severity score, Ranson score, blood lactate level on ICU admission, and length of ICU stay between patients with blood IL-6 levels ≥400 pg/ml and patients with levels <400 pg/ml. Using this PMMA-CHDF initiation criterion, PMMA-CHDF was performed on 82 SAP patients. Mean blood IL-6 level, which was 998 pg/ml on admission to the ICU, was significantly lower (335 pg/ml) after 3 days treatment of PMMA-CHDF ($p < 0.01$). In addition, heart rate, blood lactate level, and intra-abdominal pressure also decreased significantly ($p < 0.01$). At the time of weaning from PMMA-CHDF, blood IL-6 level had decreased to 99 pg/ml. The mortality rate among patients who received PMMA-CHDF was 6.1%, and significantly lower than that of patients before the introduction of PMMA-CHDF under non-renal indication (25.0%). These findings suggest that PMMA-CHDF is effective for treatment of SAP and that it can be expected to contribute to improving the outcome of SAP patients.

Severe acute pancreatitis (SAP) is a serious pathological condition requiring critical care, and its mortality rate is still high at present [1]. Regarding the pathophysiology of SAP, it was previously considered that the pancreatic enzymes spilt over into circulating blood and caused organ dysfunction. However, according to the current view, the essential characteristic of SAP is exacerbation of systemic inflammatory response syndrome (SIRS), which is triggered by local pancreatic inflammation [2].

In SIRS, various humoral mediator cascades are activated, accompanied by hypercytokinemia, which is systemic circulation of excessively produced inflammatory cytokines [3]. Hypercytokinemia has been reported to play a central role in aggravation of SAP, i.e. during progression of acute pancreatitis to SAP and the onset of organ failure induced by SAP [2, 4]. The treatment of SAP must thus focus on preventing deterioration and prolongation of SIRS, i.e. on countermeasures to deal with hypercytokinemia.

Various techniques of immunotherapy to deal with hypercytokinemia have been reported, including treatment using monoclonal antibodies to cytokines or to their receptors and using immunomodulatory drugs [5]. However, clinical effectiveness has been established for none of these methods. In our previous studies, continuous hemodiafiltration (CHDF) using a polymethyl methacrylate (PMMA) membrane hemofilter (PMMA-CHDF) was found to be capable of efficiently removing humoral mediators from blood primarily through adsorption [6–8]. In managing SAP patients, PMMA-CHDF has been applied not only as renal replacement therapy for patients complicated with acute renal failure, but also under non-renal indication for SAP patients without acute renal failure, i.e. to deal with hypercytokinemia aimed at preventing deterioration of SIRS and the onset of multiple organ failure. Such efficacy of PMMA-CHDF under non-renal indications in patients with SAP has been reported from our department [9, 10].

This paper outlines the indications for use of PMMA-CHDF and its therapeutic efficacy in patients with SAP on the basis of findings obtained in our facility.

Evaluation of the Magnitude of Hypercytokinemia in Patients with SAP

As indicated above, hypercytokinemia and activation of humoral mediator cascades, induced by acute pancreatic inflammation and subsequent local cytokine release, eventually result in SAP. Numerous studies have demonstrated that blood levels of various humoral mediators such as TNF-α, IL-1-β, IL-6, IL-10, chemokines (e.g. IL-8), and platelet activating factor are related to the severity and outcome of SAP [11–14]. These studies suggest the importance of elevated blood levels of humoral mediators in the pathophysiology of SAP. In view of the presence of many reports on the relationships of diverse cytokines to the severity and outcome of SAP, it seems unlikely that some particular cytokines are closely associated with SAP. It seems likely instead that activation of the entire cytokine network by acute pancreatic inflammation results in the development of SAP. We believe that determining the level of one of the cytokines most easily measured is a practical approach to evaluating the magnitude of activation of the entire cytokine network. We therefore introduced a real-time measurement system of IL-6 blood level, since IL-6 has several advantages over other cytokines as a biomarker of cytokine storm. IL-6 has a longer half-life than TNF-α and IL-1-β, and a reliable measuring system is available for it [15]. It has also been reported that blood IL-6 level sensitively reflects the severity of septic shock, trauma and cardiogenic shock [16]. Furthermore, blood IL-6 levels in SAP patients measured on ICU admission have been shown to correlate with APACHE II score and JMHLW (Japanese Ministry of Health, Labour, and Welfare) severity score [3]. In addition to that, although blood cytokine levels are often measured in experimental laboratories by radioimmunoassay or ELISA (enzyme-linked immunosorbent assay) using pooled samples, the technique we use for blood IL-6 measurement (rapid measurement with chemiluminescent enzyme immunoassay) can yield results in about 30 min. It is therefore clinically useful for real-time evaluation of severity and determination of therapeutic strategies for individual cases.

Criteria for Initiation of and Weaning from PMMA-CHDF

Based on the view that removal of cytokines by PMMA-CHDF is useful as a means of treating SAP, we set criteria for initiation of PMMA-CHDF and weaning from it, taking the results of the analyses presented below into account.

First, we performed a retrospective ROC analysis to evaluate the usefulness of various parameters and severity scores in predicting complications of organ failure in SAP patients. This analysis revealed that blood IL-6 level was better as a means of predicting complication by organ failure than APACHE II score, JMHLW severity score and duration (in days) of SIRS, and that the cut-off level of blood IL-6 level with the highest predictive accuracy was 400 pg/ml. Following this finding, we adopted blood IL-6 level (obtained by real-time measurement) ≥400 pg/ml as a criterion for initiation of PMMA-CHDF.

The validity of this IL-6 cut-off level was then evaluated. To this end, various severity scores, length of ICU stay, number of failed organs, and mortality were compared between 2 groups of SAP patients managed in our department between 1996 (the year of introduction of PMMA-CHDF under non-renal indication) and 2008 (n = 82), i.e. between 49 patients with blood IL-6 level ≥400 pg/ml on ICU admission and 33 patients with levels <400 pg/ml on ICU admission (table 1).

Table 1. Patient characteristics, severity scores, IL-6 blood level and mortality in all patients and comparison between groups based on IL-6 blood level on ICU admission

	All patients (n = 82)	IL-6 ≥400 pg/ml (n = 49)	IL-6 <400 pg/ml (n = 33)	p value
Age, years	50.5±17.8	49.9±18.6	51.5±16.9	n.s.
Male, n (%)	61 (74.4)	36 (73.5)	25 (75.8)	n.s.
Etiology, n				
Alcohol	40	24	16	
Gallstone	13	7	6	
Post ERCP	8	5	3	
Hypertriglyceridemia	6	3	3	
Drug induced	3	3	–	
Trauma	1	–	1	
Pancreatic tumor	1	–	1	
Idiopathic	10	7	3	
APACHE II score	12.6±6.3	14.2±6.9	11.0±6.6	<0.05 [a]
JMHLW severity score	7.1± 3.6	7.9±3.4	6.1±3.5	<0.05[a]
Ranson score	4.1±1.7	4.6±1.6	2.7±1.1	<0.001[a]
Blood IL-6 level on admission, pg/ml	998±1414	1,516±1622	198±123	<0.0001[a]
Blood lactate level on admission, mg/dl	20.0±15.5	25.7±15.2	12.3±12.1	<0.001[a]
Length of ICU stay, days	14.0±12.0	20.4±25.2	9.1±7.3	<0.05[a]
Non-survivors (mortality), n (%)	5 (6.1)	4 (8.2)	1 (3.0)	n.s.

Data are means ± SD unless stated otherwise.
[a] Unpaired Student's t test between IL-6 ≥400 and <400 pg/ml groups.

This comparison revealed significantly higher APACHE II score, JMHLW severity score, Ranson score, and blood lactate levels and significantly longer length of ICU stay in the IL-6 ≥400 pg/ml group. These findings were interpreted as supporting the validity of our blood IL-6 cut-off level (400 pg/ml).

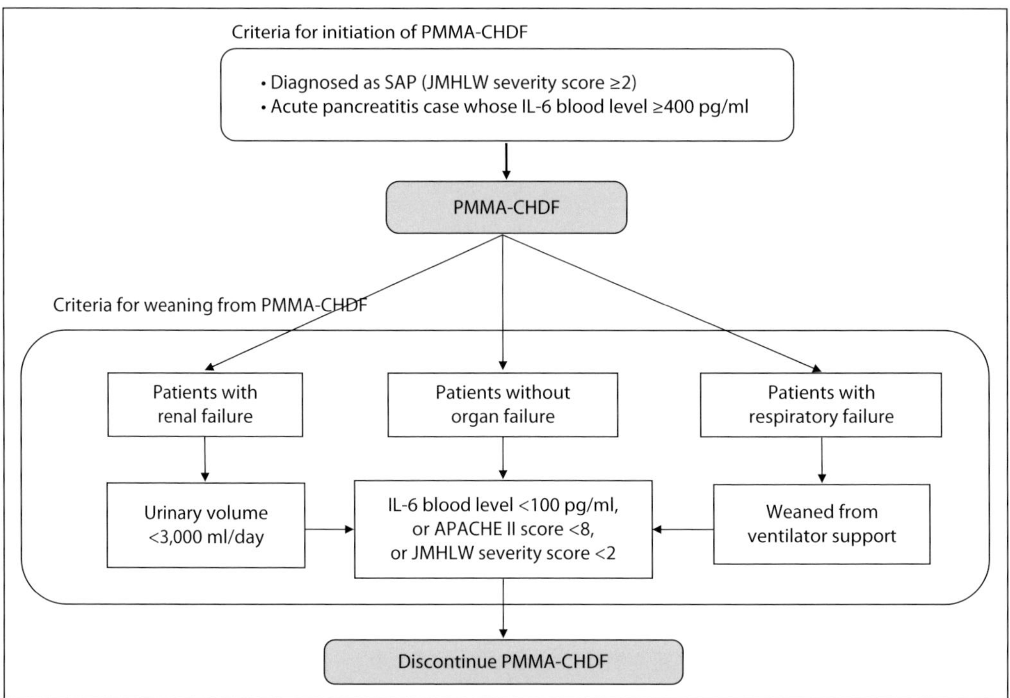

Fig. 1. Criteria for initiation of and weaning from PMMA-CHDF for acute pancreatitis patients.

On the other hand, blood IL-6 level <100 pg/ml was adopted as a criterion for weaning from PMMA-CHDF. This criterion was based on previous reports, which revealed that the efficiency of cytokine removal through adsorption onto PMMA membrane was low if blood IL-6 level was lower than 100 pg/ml [6], and that SIRS and organ failure exhibited remission in most cases with IL-6 blood level below 100 pg/ml [15].

PMMA-CHDF can be expected to play a role not only in removing cytokines as described above, but also in renal replacement therapy for patients with renal failure. Furthermore, in patients with acute respiratory distress syndrome, PMMA-CHDF has been reported to improve lung oxygenation through alleviation of interstitial lung edema as well as removal of cytokines responsible for the syndrome [17]. Taking both the efficacy of PMMA-CHDF as a means of artificial support and its efficacy as a means of cytokine removal into account, we currently use PMMA-CHDF initiation and weaning criteria composed of a combination of blood IL-6 level, severity score, presence/absence of organ failure and other factors (fig. 1).

Table 2. Comparison of heart rate, mean arterial pressure, urinary volume, intra-abdominal pressure, blood IL-6 level and blood lactate level before and after PMMA-CHDF treatment.

	Before PMMA-CHDF treatment	After PMMA-CHDF treatment for 3 days	p value
Heart rate, beats/min	120±30.2	96.4±18.0	<0.01[a]
Mean arterial pressure, mm Hg	93.7±19.5	92.7±13.3	n.s.
Urinary volume, ml/day	1,842±649	2,031±1,249	n.s.
Intra-abdominal pressure, mm Hg	12.6±5.1	10.0±5.0	<0.05[b]
Blood IL-6 level, pg/ml	998±1,414	335±463	<0.01[a]
Blood lactate level, mg/dl	20.0±15.5	6.6±3.6	<0.01[a]

Data are means ± SD.
[a] Paired Student's t test between before and after PMMA-CHDF treatment.
[b] Wicoxon's signed-rank test between before and after PMMA-CHDF treatment.

Effects of PMMA-CHDF on Blood IL-6 Level, Clinical Parameters and Outcome

Since PMMA-CHDF began to be used in our department in 1996 with the criteria noted above, 82 cases of SAP have been encountered in our department as of the year 2008. The etiology of SAP was alcoholic in 40 cases, gallstone-associated in 13 cases, post-ERCP in 8 cases, hypertriglyceridemia-associated in 6 cases, drug-induced in 3 cases, post-traumatic and tumor-associated in 1 case each and idiopathic in 10 cases. Table 1 summarizes the background variables for these patients. APACHE II score was 12.6 ± 6.3, JMHLW severity score 7.1 ± 3.6, and Ranson score 4.1 ± 1.7. All of these patients satisfied the PMMA-CHDF initiation criteria and received PMMA-CHDF for a period of 10.1 ± 10.8 days. Table 2 compares the data on various parameters after 3-day PMMA-CHDF treatment with those before PMMA-CHDF. Blood IL-6 level after 3-day PMMA-CHDF treatment (335 ± 463 pg/ml) was significantly lower than that before PMMA-CHDF (998 ± 1,414 pg/ml; table 2). This change was accompanied by significant reductions in heart rate, blood lactate level, and intra-abdominal pressure (IAP; table 2). Mean blood IL-6 level at the time of weaning from PMMA-CHDF was 99.0 pg/ml. As a result of intensive care including this treatment, 77 of the 82 patients were discharged alive from ICU, with a mortality rate of only 6.1%.

Next, we compared the outcome of these patients with those of patients treated before the introduction of PMMA-CHDF under non-renal indications (table 3). In our department, introduction of PMMA-CHDF for non-renal indications was accompanied by the introduction of selective digestive

Table 3. Comparison of mortality and incidence of infectious complications

	First period	Second period	p value
Cases, n	12	82	
Non-survivors (mortality), n (%)	3 (25.0)	5 (6.1)	<0.05[a]
Infectious complications, n (%)	6 (50.0)	9 (10.8)	<0.01[a]
Cases in need of surgical procedure, n (%)	4 (33.3)	5 (6.0)	<0.05[a]

First period: patients underwent conventional treatment. Second period: patients were treated with continuous hemodiafiltration using PMMA-CHDF and selective digestive decontamination for the prevention of bacterial translocation.
[a] χ^2 test between first and second periods.

decontamination aimed at preventing bacterial translocation. The outcome of patients before introduction of these new methods (first period, n = 12) was compared with that of patients after their introduction (second period, n = 82) for mortality, incidence of infectious complications, and number of patients requiring surgical procedures. Mortality rate differed significantly between the first period (25.0%) and the second period (6.1%, $p < 0.05$; table 3). A significant difference between the 2 periods was also noted in the incidence of infectious complications and the number of patients requiring surgical procedure, both of which were significantly lower during the second period (table 3).

These findings suggest that the mortality was reduced as a result of reduction in blood IL-6 level induced by PMMA-CHDF and accompanying reduction in heart rate, blood lactate level, and IAP.

Effects of PMMA-CHDF on Abdominal Compartment Syndrome

In patients with SAP, vascular permeability is increased by hypercytokinemia, leading to increased ascites and retroperitoneal edema [18]. Although massive intravenous fluid therapy is recommended as an initial treatment for SAP to compensate for the reduction in circulating blood volume [19, 20], massive intravenous fluid therapy in the presence of vascular endothelial hyperpermeability can induce further increases in ascites and intestinal edema, triggering elevation of IAP and the onset of abdominal compartment syndrome [21]. Abdominal compartment syndrome is a serious condition, since the elevation of IAP can induce respiratory disorder (due to elevation of the diaphragm), reduction in blood flow through abdominal organs (due to compression of abdominal vessels), and reduction of cardiac output (due to increased afterload), eventually leading to multiple organ failure [22]. Progression of organ failure due to increased IAP occasionally necessitates decompressive laparotomy [23].

To facilitate early diagnosis of abdominal compartment syndrome, we routinely measure IAP at least once a day in all cases of SAP. We previously reported that blood IL-6 level significantly correlated positively with IAP, and that the reduction in blood IL-6 level over time after the initiation of PMMA-CHDF was accompanied by significant reduction in IAP [10]. In the present study as well, we observed significant reduction in IAP after PMMA-CHDF treatment for 3 days (table 2). We also reported that the degree of change in IAP and that in blood IL-6 level demonstrated significant correlation, while no correlation was noted between the magnitude of change in IAP and cumulative water balance during the same period [10]. These findings suggest that IAP does not decrease if only water is removed, and that it does decrease if cytokines are removed followed by attenuation of inflammatory responses.

PMMA-CHDF as a Means of Cytokine Modulation

Previous studies have demonstrated that adsorption onto membrane rather than filtration or dialysis plays an important role in the removal of cytokines by PMMA-CHDF, and that this is not the case for hemofilters made of other materials such as polyacrylonitrile, polysulphone, and cellulose triacetate membrane [6, 24].

According to the recently reported results of a randomized controlled trial [25], an increase in filtrate volume used for continuous hemofiltration resulted in significant elevation of survival rate among patients with acute renal failure. Following this report, numerous studies have been conducted on high-volume hemofiltration (hemofiltration using a large volume of filtrate) and some have demonstrated that high-volume hemofiltration allowed removal of cytokines [26]. However, if blood purification aimed at removal of cytokines is performed using a hemofilter other than PMMA membrane, the filtration rate needs to be increased markedly, and blood flow therefore needs to be increased. The Japanese Guidelines on Acute Pancreatitis Management [19] recommend CHDF for patients with unstable circulation among those with SAP complicated by acute renal failure. In view of these guidelines, it does not appear useful to increase extracorporeal circulation volume in patients with unstable circulation. In this respect, application of high-volume hemofiltration in such patients is controversial. However, with PMMA-CHDF, which removes cytokines by adsorption, there is no need for such a high blood flow or filtration volume, and adequate effects can be obtained with a relatively small amount of sterile bicarbonate replacement fluid. It therefore seems useful to adopt PMMA-CHDF as a means of cytokine modulation in patients with SAP.

Conclusion

PMMA-CHDF, performed to remove cytokines and thus to modulate inflammatory reactions, rapidly diminished blood IL-6 level, accompanied by reductions in heart rate, blood lactate level, and IAP. The mortality rate of patients after the introduction of PMMA-CHDF under non-renal indication was 6.1%, and it was significantly lower than that before its introduction. These findings suggest that PMMA-CHDF is effective for treatment of SAP and has strong potential to contribute to improving the outcome of SAP patients. A large-scale randomized controlled trial needs to be carried out to obtain further evidence for its efficacy.

References

1 Swaroop VS, Chari ST, Clain JE: Severe acute pancreatitis. JAMA 2004;291:2865–2868.
2 Bhatia M, Brady M, Shokuhi S, Christmas S, Neoptolemos JP, Slavin J: Inflammatory mediators in acute pancreatitis. J Pathol 2000;190:117–125.
3 Hirasawa H, Oda S, Matsuda K, Watanabe E: Basic concept and definition of SIRS and sepsis: present consideration and future perspectives (in Japanese). Nippon Rinsho 2004;62:2177–2183.
4 Mofidi R, Duff MD, Wigmore SJ, Madhavan KK, Garden OJ, Parks RW: Association between early systemic inflammatory response, severity of multiorgan dysfunction and death in acute pancreatitis. Br J Surg 2006;93:738–744.
5 Zidek Z, Anzenbacher P, Kmonickova E: Current status and challenges of cytokine pharmacology. Br J Pharmacol 2009;157:342–361.
6 Matsuda K, Hirasawa H, Oda S, Shiga H, Nakanishi K: Current topics on cytokine removal technologies. Ther Apher 2001;5:306–314.
7 Nakada TA, Hirasawa H, Oda S, Shiga H, Matsuda K: Blood purification for hypercytokinemia. Transfus Apher Sci 2006;35:253–264.
8 Hirasawa H, Oda S, Matsuda K: Continuous hemodiafiltration with cytokine-adsorbing hemofilter in the treatment of severe sepsis and septic shock. Contrib Nephrol 2007;156:365–370.
9 Moriguchi T, Hirasawa H, Oda S, Shiga H, Nakanishi K, Matsuda K, Nakamura M, Yokohari K, Hirano T, Hirayama Y, Watanabe E: A patient with severe acute pancreatitis successfully treated with a new critical care procedure. Ther Apher 2002;6:221–224.
10 Oda S, Hirasawa H, Shiga H, Matsuda K, Nakamura M, Watanabe E, Moriguchi T: Management of intra-abdominal hypertension in patients with severe acute pancreatitis with continuous hemodiafiltration using a polymethyl methacrylate membrane hemofilter. Ther Apher Dial 2005;9:355–361.
11 Gukovskaya AS, Gukovsky I, Zaninovic V, Song M, Sandoval D, Gukovsky S, Pandol SJ: Pancreatic acinar cells produce, release, and respond to tumor necrosis factor-alpha: role in regulating cell death and pancreatitis. J Clin Invest 1997;100:1853–1862.
12 Mayer J, Rau B, Gansauge F, Beger HG: Inflammatory mediators in human acute pancreatitis: clinical and pathophysiological implications. Gut 2000;47:546–552.
13 Shimada M, Andoh A, Hata K, Tasaki K, Araki Y, Fujiyama Y, Bamba T: IL-6 secretion by human pancreatic periacinar myofibroblasts in response to inflammatory mediators. J Immunol 2002;168:861–868.
14 Granger J, Remick D: Acute pancreatitis: models, markers, and mediators. Shock 2005;24(suppl 1):45–51.

15 Oda S, Hirasawa H, Shiga H, Nakanishi K, Matsuda K, Nakamua M: Sequential measurement of IL-6 blood levels in patients with systemic inflammatory response syndrome (SIRS)/sepsis. Cytokine 2005;29:169–175.
16 Martin C, Boisson C, Haccoun M, Thomachot L, Mege JL: Patterns of cytokine evolution (tumor necrosis factor-alpha and interleukin-6) after septic shock, hemorrhagic shock, and severe trauma. Crit Care Med 1997;25:1813–1819.
17 Hirayama Y, Hirasawa H, Oda S, Shiga H, Matsuda K, Ueno H, Nakamura M, Moriguchi T, Watanabe E, Nitta M, Abe R, Nakada T, Kobe Y, Tataishi Y: Indication and clinical efficacy of continuous hemodiafiltration (CHDF) for ARDS (in Japanese). ICU CCU 2004;28:S122–S124.
18 Raraty MG, Connor S, Criddle DN, Sutton R, Neoptolemos JP: Acute pancreatitis and organ failure: pathophysiology, natural history, and management strategies. Curr Gastroenterol Rep 2004;6:99–103.
19 Takeda K, Takada T, Kawarada Y, Hirata K, Mayumi T, Yoshida M, Sekimoto M, Hirota M, Kimura Y, Isaji S, Koizumi M, Otsuki M, Matsuno S: JPN guidelines for the management of acute pancreatitis: medical management of acute pancreatitis. J Hepatobiliary Pancreat Surg 2006;13:42–47.
20 Frossard JL, Steer ML, Pastor CM: Acute pancreatitis. Lancet 2008;371:143–152.
21 Al-Bahrani AZ, Abid GH, Holt A, McCloy RF, Benson J, Eddleston J, Ammori BJ: Clinical relevance of intra-abdominal hypertension in patients with severe acute pancreatitis. Pancreas 2008;36:39–43.
22 An G, West MA: Abdominal compartment syndrome: a concise clinical review. Crit Care Med 2008;36:1304–1310.
23 Cheatham ML, Malbrain ML, Kirkpatrick A, Sugrue M, Parr M, De Waele J, Balogh Z, Leppaniemi A, Olvera C, Ivatury R, D'Amours S, Wendon J, Hillman K, Wilmer A: Results from the International Conference of Experts on Intra-abdominal Hypertension and Abdominal Compartment Syndrome. II. Recommendations. Intensive Care Med 2007;33:951–962.
24 Matsuda K, Moriguchi T, Harii N, Goto J: Comparison of efficacy between continuous hemodiafiltration with a PMMA membrane hemofilter and a PAN membrane hemofilter in the treatment of a patient with septic acute renal failure. Transfus Apher Sci 2009;40:49–53.
25 Ronco C, Bellomo R, Homel P, Brendolan A, Dan M, Piccinni P, La Greca G: Effects of different doses in continuous veno-venous haemofiltration on outcomes of acute renal failure: a prospective randomised trial. Lancet 2000;356:26–30.
26 Cole L, Bellomo R, Journois D, Davenport P, Baldwin I, Tipping P: High-volume haemofiltration in human septic shock. Intensive Care Med 2001;27:978–986.

Ryuzo Abe
Department of Emergency and Critical Care Medicine, Chiba University Graduate School of Medicine
1-8-1 Inohana Chuo
Chiba 260-8677 (Japan)
Tel. +81 43 222 7171, Fax +81 43 226 2371, E-Mail ryuzo@pf6.so-net.ne.jp

Suzuki H, Hirasawa H (eds): Acute Blood Purification.
Contrib Nephrol. Basel, Karger, 2010, vol 166, pp 64–72

Blood Purification in Fulminant Hepatic Failure

Koichiro Shinozaki · Shigeto Oda · Ryuzo Abe · Yoshihisa Tateishi · Takehito Yokoi · Hiroyuki Hirasawa

Department of Emergency and Critical Care Medicine, Chiba University Graduate School of Medicine, Chiba, Japan

Abstract

Fulminant hepatic failure (FHF) can be described as a potentially fatal condition presenting with hepatic encephalopathy (HE) and coagulopathy associated with acute hepatic dysfunction, regardless of its etiology. Blood purification (BP) is expected to be effective against HE and coagulopathy in FHF. In this paper, we outline the objectives and methods of BP in the treatments of cases with FHF and indicate a concrete method for and outcomes of BP at our facility. In high-flow dialysate continuous hemodiafiltration (HFCHDF), the conventional CHDF bedside console is connected to a personal dialysis console to induce a high flow rate of dialysate. With this method, the dialysate flow rate is about 500 ml/min at maximum, equivalent to about 50 times the dialysate flow rate during ordinary CHDF. The role of plasma exchange (PE) is considered a means of replacing useful substances, such as clotting factors in fresh frozen plasma rather than a means of removing pathogenic substances. As needed, slow PE(SPE) can be incorporated by connection in series. Analysis of data from 90 patients with FHF who underwent BP at our facility after 1990 revealed that restoration of consciousness was achieved in 33 (70.2%) of 47 cases when treated with HFCHDF. This survival in the HFCHDF group was significantly higher than that in the CHDF group. Analysis of data from cases in which ammonia could be measured continuously revealed that blood ammonia level decreased over time following HFCHDF. We also revealed that HFCHDF was useful for preventing the side effects of PE, such as hypernatremia, metabolic alkalosis, and sharp decrease in colloid osmotic pressure. It is concluded that HFCHDF is useful in the treatment of HE and for preventing the side effects of PE. Therefore, we suggested that HFCHDF + SPE should be standardized for the treatment of FHF.

Acute liver failure (ALF) is a condition characterized by acute hepatic dysfunction which has been developed within 26 weeks after the initial etiological event and presents with disturbance of consciousness and coagulopathy (PT-INR ≥1.5) in patients without a history of liver disease such as cirrhosis [1]. Both its prognosis and treatment vary markedly depending on its etiology [1]. Fulminant hepatic failure (FHF) is a type of ALF with rapid aggravation involving intense hepatocellular necrosis and featuring a high mortality [2]. Although no clear definition of FHF is available, since the study by Trey et al. [2] many reports [3–5] have been published in which FHF was defined as a type of ALF which has been developed less than 8 weeks following the initial etiological event and which involves stage 2 or more severe hepatic encephalopathy (HE) [6], i.e. disturbance of consciousness ranging from lethargy or inappropriate behavior to coma. In Japan, coagulopathy with PT activity no higher than 40% is considered another condition related to definition of FHF [7].

FHF can be briefly described as a potentially fatal condition presenting with HE and coagulopathy associated with acute hepatic dysfunction, regardless of its etiology. Blood purification (BP) is expected to be effective against HE and coagulopathy in FHF. In this paper, we outline the objectives and methods of BP in the treatment of the cases with FHF and indicate a concrete method for and outcomes of BP at our facility.

Purpose of Blood Purification

Therapeutic interventions aimed at inhibiting aggravation of FHF and reversing deteriorated hepatic function should be selected depending on etiology in individual cases [1]. BP is, as a rule, performed not as a means of eliminating etiological factors, and instead aims to avoid progression of conditions such as HE and coagulopathy (arising from hepatic failure) and compensating for them. Therefore, the purpose of BP is as an artificial support for minimal liver function required to sustain the life of the patient.

The basic strategies for treatment of FHF can be divided into 2 types. In cases in which hepatic dysfunction is considered irreversible, liver transplantation [8] is performed, while in cases in which hepatic dysfunction is reversible, liver cell regeneration [9] is anticipated. BP is used as a bridge to maintain the patient's life and maintain his/her general condition until support of liver function is no longer required. It is performed as a bridge to either transplantation or regeneration.

The compensatory functions and other roles of BP involve: (1) removal of materials such as those causing HE; (2) replacement of substances such as clotting factors; (3) correction of water, electrolyte, and acid-base balance in patients with acute renal failure [10], a common complication of FHF, and (4) removal of various pro-inflammatory cytokines believed to elevate intracranial pressure and participate in the mechanism of onset of HE [11, 12]. Since pro-inflammatory cytokines are involved in the pathophysiology of systemic inflammatory response syndrome (SIRS) [13] and since SIRS is suggestive of a poor prognosis for FHF [13], removal of pro-inflammatory cytokines is important during BP for FHF. The characteristics of various techniques of BP in patients with FHF will be outlined below, focusing on these 4 features.

Hemodialysis and Hemofiltration

Hemodialysis is a method of treatment involving substance removal across the semi-permeable membrane of the dialyzer, and it aims to purify blood by removing hazardous substances primarily through diffusion. If hemodialysis is performed continuously, it is termed continuous hemodialysis (CHD), to distinguish it from intermittent treatment. In the treatment of cases with ALF, CHD is recommended [1, 14]. CHD is an excellent means of removing low-molecular-weight substances and is usually applied to cases of FHF to remove and adjust electrolytes or metabolites when FHF is complicated by renal failure [1].

Ammonia, which appears to be associated with increased intracranial pressure [15], is considered one of the substances responsible for onset of HE [6]. Removal of ammonia is thus thought to contribute to alleviation of the symptoms of HE. Ammonia can be removed by hemodialysis. Denis et al. [5] reported that, of 39 patients with FHF who underwent hemodialysis with polyacrylonitrile membrane, 24 (61.5%) regained consciousness. Following their report, hemodialysis has been utilized as a useful means of treating HE.

Hemofiltration is a method of treatment designed to remove hazardous substances from the blood by ultrafiltration alone, using a hemofilter without dialysate. Hemofiltration is superior to hemodialysis in efficiency of removing substances of intermediate molecular weight. It has also been reported that hemofiltration at filtration rates of 35 ml/min or above is capable of removing cytokines from blood [16] and that it improves the survival of patients [17]. These findings can be explained by the hypothesis that high filtration rates are needed to remove various pro-inflammatory cytokines of molecular weights about 20,000 Daltons in hemofiltration. As noted above, the prognosis of FHF is closely related to SIRS. It is thus very likely that various pro-inflammatory cytokines [11, 13, 18] are closely associated with the etiology and prognosis of FHF. It therefore seems essential to plan removal of pro-inflammatory cytokines when treating patients with FHF by BP. It was recently reported that SIRS is closely involved in the renal dysfunction observed in patients with FHF [18]. A common etiological factor appears to be present in both the renal dysfunction associated with sepsis and the mechanism of onset of renal dysfunction in patients with FHF [18].

For the reasons given above, it is desirable to apply BP continuously using a combination of hemodialysis and hemofiltration when dealing with patients with FHF regardless of the presence or absence of complication of renal failure. Continuous hemodiafiltration (CHDF) should thus be selected for such cases.

In the past, we used BP involving a combination of slow plasma exchange (SPE) and CHDF (SPE + CHDF), as described in detail below, when treating the patients with FHF. However, some patients with FHF failed to regain consciousness. In such cases, we attempted in exploratory fashion to increase the quantity of BP. This resulted in marked improvement in level of consciousness. Details of the methods we adopted to increase the quantity of BP will be provided below. We tried various methods of increasing the quantity of BP, and found that a method involving increase dialysate flow in CHDF has fewer adverse effects on the circulation and is more desirable in terms of cost-effectiveness. Since 1996, the dialysate flow rate during ordinary CHDF at our facility has been set at 500 ml/min, about 50 times the conventional level, and BP has been performed as a combination of high-flow dialysate CHDF (HFCHDF) and SPE at our facility, yielding favorable results [19].

Plasmapheresis

Plasmapheresis is a method of treatment in which the plasma components separated with a plasma separator are subjected to plasma exchange (PE), plasma adsorption, double-filtration plasmapheresis with a secondary membrane, and other treatments. Since the report by Lepore et al. [20] on their experience with PE in 5 cases of FHF, PE has been used as a common technique of plasmapheresis in cases of FHF.

PE is a method of BP designed to achieve therapeutic efficacy through discarding the separated patient's plasma and replacing plasma from healthy donors (often fresh frozen plasma, FFP). It is capable of removing hazardous substances of low to high molecular weight and is additionally capable of replacing useful substances deficient in the patient's blood. When treating the patients with FHF, PE is performed to remove substances responsible for HE and various pro-inflammatory cytokines [21] and also to replenish clotting factors through replacement of the patient's plasma with FFP. However, this method features several problems, including the consumption of large quantities of FFP, a valuable medical resource, as well as possible side effects arising from such massive replacement with FFP, such as hypernatremia, metabolic alkalosis, sharp decrease in colloid osmotic pressure, and others [22]. For these reasons, it is not desirable to replace large amounts of plasma rapidly for the purpose of compensation for hepatic function. Based on assessment of the capability of HFCHDF to remove substances responsible for the onset of FHF, a view currently prevailing in Japan concerning the role of PE is that it should be considered a means of replacing useful substances such as clotting factors rather than a means of removing pathogenic substances. Therefore, BP for FHF is now frequently performed with PE + HFCHDF with addition of PE, if needed, to the basic method of HFCHDF [19, 23–25]. This approach is also useful for preventing the side effects noted above [22].

Molecular Adsorbents Recirculating System

In controlled studies, Heemann et al. [26] demonstrated the possibility that a molecular adsorbents recirculating system (MARS) can increase both survival and restoration of consciousness in patients with HE. MARS is a system by which whole blood of the patient is dialyzed with dialysate containing high concentrations of albumin and a high-flux membrane dialyzer, followed by dialysis of albumin-containing dialysate with conventional dialysate and its regeneration through passage through an adsorbing column. A question has been raised concerning the feasibility of removing substances responsible for HE through removal of albumin alone with this system. Furthermore, some investigators have recommended use of high-flow hemodialysis to reduce blood levels of ammonia satisfactorily, since ammonia-removing efficiency is low with MARS [27].

Bioartificial Liver Devices

In contrast to hemodiafiltration and plasmapheresis which serve as non-biological artificial liver support, the term 'biological artificial liver support' is used to indicate the method of BP which incorporates homogeneous or heterogeneous liver cells into the bioreactors and blood is purified by hemoperfusion during blood oxygenation with the oxygenator. The therapeutic efficacy of this method of BP depends on the properties of the cultured cells incorporated into the bioartificial liver device. In the controlled trials conducted to date, the usefulness of this method was demonstrated only when it was performed with the device adopting porcine hepatocytes in the study reported by Demetriou et al. [28].

A key factor determining the success in research on biological artificial liver is establishment of a cell line. The cell line for use in biological artificial liver is required to be stable and functional. One way of resolving the poor functionality is the hybrid approach by which the poor function is made up for by the above-mentioned non-biological artificial liver. In any event, this kind of research may lead to regeneration of liver cells and its marked advances from now on are expected.

Major methods of BP currently used have been outlined above, referring to reports from Western countries as well. The concrete method and results of non-biological artificial liver, playing a central role in FHF management, at our facility are presented below.

Clinical Implementation of Blood Purification in Our Group

At our facility, PE had been used independently as an artificial liver in cases of FHF until 1992. In 1992, SPE + CHDF began to be used to prevent complications

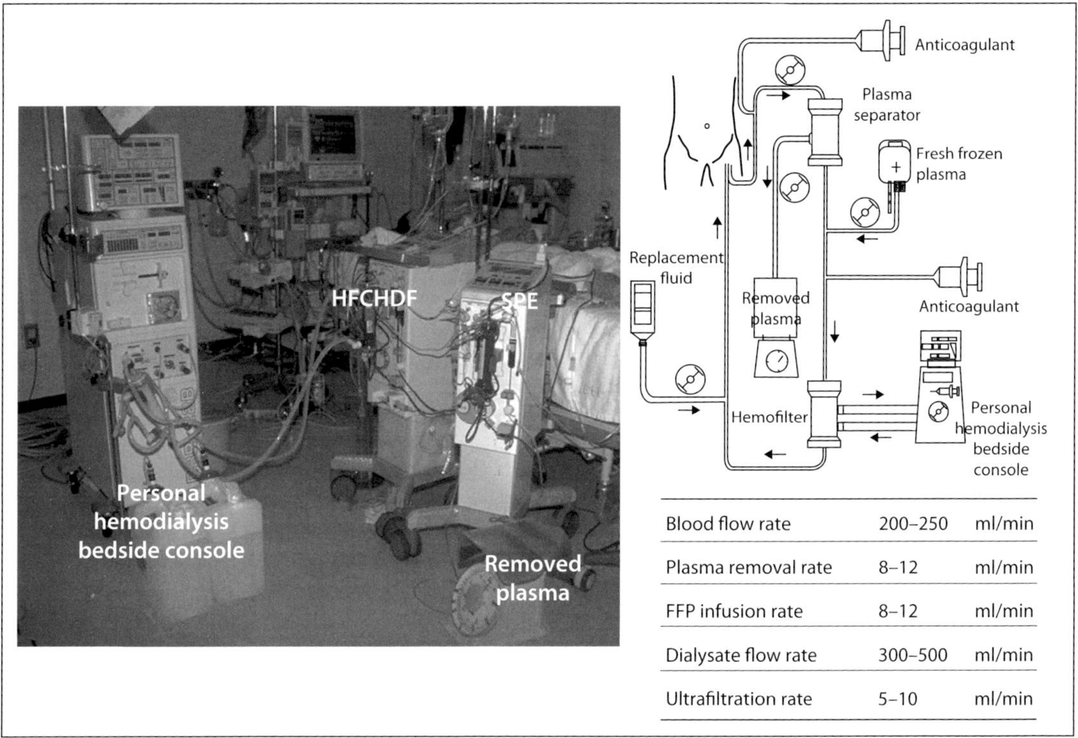

Blood flow rate	200–250	ml/min
Plasma removal rate	8–12	ml/min
FFP infusion rate	8–12	ml/min
Dialysate flow rate	300–500	ml/min
Ultrafiltration rate	5–10	ml/min

Fig. 1. Flow diagram and operating condition of SPE + HFCHDF. The image shows a patient undergoing this treatment in our department.

arising from the treatment using PE alone (e.g. hypernatremia, metabolic alkalosis, sudden reduction in colloid osmotic pressure), and to achieve efficient removal of substances of low to intermediate molecular weight. This method involves SPE (6–8 h of PE with the substitution volume set at 1 plasma volume) connected in series to CHDF. However, as noted above, we subsequently became aware of the need for a more powerful BP method to remove HE-causing substances. After assessing various methods for increasing the quantity of BP rate, we switched from CHDF to HFCHDF in 1996.

With HFCHDF, the conventional CHDF bedside console is connected to a personal dialysis bedside console to induce high dialysate flow dialysis. Furthermore, vascular access is modified from the conventional side hole type to the end hole type, to ensure the high blood flow needed for a high flow rate of dialysis fluid. With this method, the dialysate flow rate is about 500 ml/min at maximum and the blood flow is about 250 ml/min at maximum. As needed, SPE can be incorporated by connection in series. After completion of SPE, HFCHDF alone is continued for the rest of the day (fig. 1). A hemofilter made

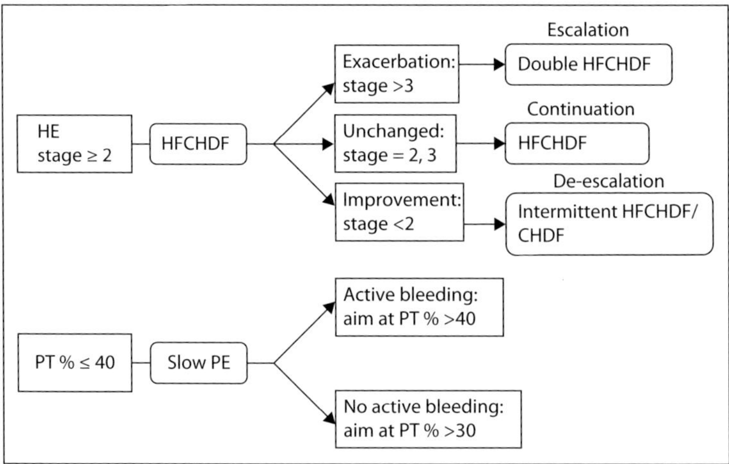

Fig. 2. Indications for HFCHDF and SPE. We now make it a rule to apply HFCHDF when treating patients with stage 2 or more severe HE. If HE progresses to a stage >3 despite such treatment, we secure 2 series of vascular access (the femoral vein and the internal carotid vein) and perform HFCHDF in duplicate fashion. If HE improves to a stage <2, the duration of HFCHDF per day is shortened or the quantity of blood purification is reduced (with switching from HFCHDF to conventional CHDF). SPE is applied as needed for patients exhibiting a tendency toward active bleeding, with the goal set at PT% >40, and patients without bleeding tendency (goal: PT% >30).

from polymethyl methacrylate membrane, which is primarily used for the treatment of sepsis in Japan, can additionally adsorb cytokines. The use of the hemofilter made from polymethyl methacrylate membrane in combination with about 5 ml/min hemofiltration has been reported to yield cytokine-removing effects useful in the treatment of sepsis [29, 30]. We therefore use polymethyl methacrylate membrane hemofilter for HFCHDF, setting the filtration flow rate at 5–10 ml/min.

After repeated modifications over time, we now make it a rule to apply HFCHDF when treating patients with stage 2 or more severe HE (fig. 2). If HE progresses to a stage >3 despite such treatment, we secure 2 series of vascular access (the femoral vein and the internal carotid vein) and perform HFCHDF in duplicate fashion. If HE improves to a stage <2, the duration of HF per day is shortened or the quantity of BP is reduced (with switching of HFCHDF to conventional CHDF). SPE is applied as needed for patients exhibiting a tendency toward active bleeding, with the goal set at PT% >40, and patients without bleeding tendency (goal: PT% >30). As a rule, SPE is performed for 1 session per day, with the plasma substitution volume set at 50 ml/kg and a duration of 6–8 h, as described above.

Analysis of data from 90 patients with FHF who underwent BP at our facility after 1990 revealed that restoration of consciousness was achieved in 33 (70.2%)

of 47 cases after HFCHDF [19]. This rate for the HFCHDF group was significantly higher than that for the SPE + CHDF group and that for the uncombined HE group, indicating the usefulness of HFCHDF in the treatment of HE. Analysis of data from cases in which ammonia could be measured continuously revealed that blood ammonia level decreased over time following HFCHDF [19].

Conclusions

The method of BP performed at our facility for cases of FHF has been indicated in this paper, with reference to our therapeutic strategy. FHF is a fatal disease, the features of which vary markedly among individual cases. When considering methods of management of diseases with unique features like FHF, it is difficult to demonstrate the usefulness of a single treatment approach that is applicable to all cases. Furthermore, healthcare environments and ethical perspectives vary from country to country. It therefore seems useful to adopt an experience-based approach to discussion of the validity of methods for treatment of FHF. It is essential to perform BP based on well-defined management goals and to select a method and quantity of BP optimal for the condition of a given patient.

References

1 Polson J, Lee WM: AASLD position paper: the management of acute liver failure. Hepatology 2005;41:1179–1197.
2 Trey C, Lipworth L, Chalmers TC, et al: Fulminant hepatic failure: presumable contribution to halothane. N Engl J Med 1968; 279:798–801.
3 Mas A, Rodes J: Fulminant hepatic failure. Lancet 1997;349:1081–1085.
4 Buckner CD, Clift RA, Volwiler W, et al: Plasma exchange in patients with fulminant hepatic failure. Arch Intern Med 1973;132: 487–492.
5 Denis J, Opolon P, Nusinovici V, et al: Treatment of encephalopathy during fulminant hepatic failure by haemodialysis with high permeability membrane. Gut 1978;19: 787–793.
6 Riordan SM, Williams R: Treatment of hepatic encephalopathy. N Engl J Med 1997; 337:473–479.
7 Mochida S, Fujiwara K: Symposium on clinical aspects in hepatitis virus infection: recent advances in acute and fulminant hepatitis in Japan. Intern Med 2001;40:175–177.
8 O'Leary JG, Lepe R, Davis GL: Indications for liver transplantation. Gastroenterology 2008;134:1764–1776.
9 Fausto N, Campbell JS, Riehle KJ: Liver regeneration. Hepatology 2006;43:S45–S53.
10 Ring-Larsen H, Palazzo U: Renal failure in fulminant hepatic failure and terminal cirrhosis: a comparison between incidence, types, and prognosis. Gut 1981;22:585–591.
11 Jalan R, Pollok A, Shah SH, et al: Liver derived pro-inflammatory cytokines may be important in producing intracranial hypertension in acute liver failure. J Hepatol 2002;37:536–538.
12 Odeh M: Pathogenesis of hepatic encephalopathy: the tumour necrosis factor-alpha theory. Eur J Clin Invest 2007;37:291–304.
13 Rolando N, Wade J, Davalos M, et al: The systemic inflammatory response syndrome in acute liver failure. Hepatology 2000;32: 734–739.

14 Davenport A, Will EJ, Davidson AM: Improved cardiovascular stability during continuous modes of renal replacement therapy in critically ill patients with acute hepatic and renal failure. Crit Care Med 1993;21: 328–338.
15 Bernal W, Hall C, Karvellas CJ, et al: Arterial ammonia and clinical risk factors for encephalopathy and intracranial hypertension in acute liver failure. Hepatology 2007;46:1844–1852.
16 Ronco C, Tetta C, Mariano F, et al: Interpreting the mechanisms of continuous renal replacement therapy in sepsis: the peak concentration hypothesis. Artif Organs 2003; 27:792–801.
17 Ronco C, Bellomo R, Homel P, et al: Effects of different doses in continuous veno-venous haemofiltration on outcomes of acute renal failure: a prospective randomised trial. Lancet 2000;356:26–30.
18 Leithead JA, Ferguson JW, Bates CM, et al: The systemic inflammatory response syndrome is predictive of renal dysfunction in patients with non-paracetamol-induced acute liver failure. Gut 2009;58:443–449.
19 Yokoi T, Oda S, Shiga H, et al: Efficacy of high-flow dialysate continuous hemodiafiltration in the treatment of fulminant hepatic failure. Transfus Apher Sci 2009;40:61–70.
20 Lepore MJ, Martel AJ: Plasmapheresis with plasma exchange in hepatic coma: methods and results in five patients with acute fulminant hepatic necrosis. Ann Intern Med 1970; 72:165–174.
21 Iwai H, Nagaki M, Naito T, et al: Removal of endotoxin and cytokines by plasma exchange in patients with acute hepatic failure. Crit Care Med 1998;26:873–876.
22 Sadahiro T, Hirasawa H, Oda S, et al: Usefulness of plasma exchange plus continuous hemodiafiltration to reduce adverse effects associated with plasma exchange in patients with acute liver failure. Crit Care Med 2001;29:1386–1392.
23 Yonekawa C, Nakae H, Tajimi K, et al: Effectiveness of combining plasma exchange and continuous hemodiafiltration in patients with postoperative liver failure. Artif Organs 2005;29:324–328.
24 Sadamori H, Yagi T, Inagaki M, et al: High-flow-rate haemodiafiltration as a brain-support therapy proceeding to liver transplantation for hyperacute fulminant hepatic failure. Eur J Gastroenterol Hepatol 2002;14:435–439.
25 Yoshiba M, Inoue K, Sekiyama K, et al: Favorable effect of new artificial liver support on survival of patients with fulminant hepatic failure. Artif Organs 1996;20:1169–1172.
26 Heemann U, Treichel U, Loock J, et al: Albumin dialysis in cirrhosis with superimposed acute liver injury: a prospective, controlled study. Hepatology 2002;36:949–958.
27 Ferenci P, Kramer L: MARS and the failing liver: Any help from the outer space? Hepatology 2007;46:1682–1684.
28 Demetriou AA, Brown RS Jr, Busuttil RW, et al: Prospective, randomized, multicenter, controlled trial of a bioartificial liver in treating acute liver failure. Ann Surg 2004;239: 660–667.
29 Hirasawa H, Oda S, Matsuda K: Continuous hemodiafiltration with cytokine-adsorbing hemofilter in the treatment of severe sepsis and septic shock. Contrib Nephrol 2007;156: 365–370.
30 Nakada TA, Oda S, Matsuda K, et al: Continuous hemodiafiltration with PMMA Hemofilter in the treatment of patients with septic shock. Mol Med 2008;14:257–263.

Koichiro Shinozaki
Department of Emergency and Critical Care Medicine, Chiba University Graduate School of Medicine
1-8-1 Inohana, Chuo-ku
Chiba City, Chiba, 260-8677 (Japan)
Tel. +81 43 226 2341, Fax +81 43 226 2371, E-Mail shino@gk9.so-net.ne.jp

Suzuki H, Hirasawa H (eds): Acute Blood Purification.
Contrib Nephrol. Basel, Karger, 2010, vol 166, pp 73–82

Treatment of Severe Sepsis and Septic Shock by CHDF Using a PMMA Membrane Hemofilter as a Cytokine Modulator

Masataka Nakamura · Shigeto Oda · Tomohito Sadahiro · Yoh Hirayama · Eizo Watanabe · Yoshihisa Tateishi · Taka-aki Nakada · Hiroyuki Hirasawa

Department of Emergency and Critical Care Medicine, Chiba University Graduate School of Medicine, Chiba, Japan

Abstract

It has been reported that various types of blood purification intended for the removal of humoral mediators, such as cytokines, were performed in patients with severe sepsis/septic shock. While high-volume hemofiltration, hemofiltration using high cut-off membrane filters, and direct hemoperfusion with a polymyxin-B immobilized column are widely used in the treatment of severe sepsis/septic shock, we perform continuous hemodiafiltration using a polymethylmethacrylate membrane hemofilter (PMMA-CHDF), which shows an excellent cytokine-adsorbing capacity, for the treatment of severe sepsis/septic shock. In our previous study, it was found that PMMA-CHDF could efficiently remove various pro-inflammatory cytokines such as TNFα, IL-6 and IL-8 from the bloodstream, resulting in early recovery from septic shock. Furthermore, PMMA-CHDF could remove anti-inflammatory cytokines such as IL-10 from bloodstream, suggesting that it might improve immunoparalysis as well. These findings suggest that PMMA-CHDF is useful for the treatment of patients with severe sepsis/septic shock as a cytokine modulator.

Despite the recent advances in critical care medicine, the mortality of severe sepsis/septic shock still remains high [1]. It has been reported that pro-inflammatory cytokines play a pivotal role in the early stage of severe sepsis/septic shock, and that overproduction of pro-inflammatory cytokines is known to cause shock and organ failure [2]. It has been also reported that anti-inflammatory cytokines are predominant in the late stage of severe sepsis/septic shock, and that their overproduction stimulates disease progression from compensatory anti-inflammatory response syndrome to immunoparalysis [3]. Recently, Ronco et al. [4] have proposed that, in severe sepsis/septic shock, immunohomeostasis breaks down due to imbalance of pro- and anti-inflammatory cytokines, both of which are produced excessively. Restoration of immunohomeostasis with various agents, either immunostimulants or immunosuppressants, has been attempted in immunomodulatory therapy in patients with severe sepsis/septic shock [5, 6].

Continuous hemofiltration has also been performed as one of the immunomodulatory therapies in patients with severe sepsis/septic shock [7]. Continuous hemofiltration, which can remove various intermediate-molecular-weight substances (30–40 kDa) from the bloodstream, has been expected to remove various cytokines (5–30 kDa) non-selectively. Therefore, continuous hemofiltration was performed as a cytokine modulator in the treatment of patients with severe sepsis/septic shock [7]. However, whether continuous hemofiltration could actually remove various cytokines was controversial for a long period [8]. On the other hand, it has recently been recognized that additional technical innovations to conventional hemofiltration could remove various cytokines effectively. More specifically, high-volume hemofiltration (HVHF) involving enhanced ultrafiltration flow [9] and hemofiltration using high cutoff membrane hemofilters [10] have recently been widely performed in patients with septic shock in western countries, aiming at cytokine removal and early recovery from shock [9, 10].

While HVHF or hemofiltration with high cutoff membrane hemofilter were being widely used as cytokine modulators, we performed continuous hemodiafiltration using a polymethylmethacrylate membrane hemofilter (PMMA-CHDF) for various cytokine-associated disorders, such as severe acute pancreatitis, acute respiratory distress syndrome and severe sepsis/septic shock [11–13], since the PMMA membrane hemofilter adsorbs a wide variety of cytokines [11]. In this article, we discuss the clinical efficacy of PMMA-CHDF intended for cytokine removal in the treatment of severe sepsis/septic shock based on our own experience.

Procedures in PMMA-CHDF for the Treatment of Severe Sepsis/Septic Shock

In our institution, the blood IL-6 level is routinely measured as a biomarker of cytokine storms in all patients admitted to the ICU. Our previous study revealed that blood IL-6 levels increased with exacerbation of sepsis (i.e. progression of sepsis to severe sepsis and septic shock) [14]. Therefore, in the treatment of patients with severe sepsis/septic shock, the indication of PMMA-CHDF for the purpose of cytokine removal is determined according to the blood IL-6 level. Since we found that a persistently high blood IL-6 level (>1,000 pg/ml) was associated with a risk of progression to multiple organ dysfunction syndrome

[11, 13, 14], we defined the cutoff point for blood IL-6 level as 1,000 pg/ml for initiation or termination of PMMA-CHDF.

The procedures for PMMA-CHDF are outlined below. Vascular access is provided in the usual venovenous mode. The PMMA membrane hemofilter that we usually use has a membrane area of 1.0 m^2 and a standard pore size (Hemofeel CH1.0, Toray Medical, Japan). Operating conditions for PMMA-CHDF are as follows: blood flow rate 80–120 ml/min, dialysate flow rate 500–1,000 ml/h, and filtration rate 300–500 ml/h. Since PMMA-CHDF removes cytokines by adsorption to the PMMA membrane and not by convection or diffusion [11, 12], no particular increase in dialysate flow rate or filtration rate is required. Our operating condition of PMMA-CHDF for cytokine removal is similar to those employed for routine renal support in most institutions. Thus, PMMA-CHDF can be performed gently, continuously and easily with hemodynamically unstable septic patients.

Cytokine-Removing Effects of PMMA-CHDF

The capacity of PMMA-CHDF to adsorb cytokines is an important issue. We previously investigated the changes in blood levels of various cytokines during PMMA-CHDF therapy for 3 days in critically ill patients, including those with sepsis (fig. 1) [11]. Patients were divided into 2 groups based on the blood levels of the cytokine of interest before the initiation of PMMA-CHDF, a 'high-level' and a 'low-level' group, and the changes in blood cytokine levels during PMMA-CHDF were compared. Significant decreases in blood levels of both pro- (i.e. TNFα, IL-6 and IL-8) and anti-inflammatory (i.e. IL-10) cytokines were observed in the each 'high-level' group. This demonstrated that PMMA-CHDF efficiently removed both pro- and anti-inflammatory cytokines from the bloodstream. It was also found that no such non-selective decrease in blood levels of various cytokines was observed during CHDF using a hemofilter composed of other membranes (i.e. ethylene vinyl alcohol, polysulfone or polyacrylonitrile) when CHDF was performed under the same operational conditions as PMMA-CHDF [11]. Thus, the decrease in blood levels of cytokines through adsorption was a unique characteristic of PMMA-CHDF.

Therapeutic Effects of Early Implementation of PMMA-CHDF in Septic Shock

In this section, we discuss whether PMMA-CHDF is effective in treating cytokine storm, a major pathophysiological aspect of septic shock, and improving dysoxia in patients with septic shock, as has been reported for HVHF performed for the treatment of septic shock in western countries [9].

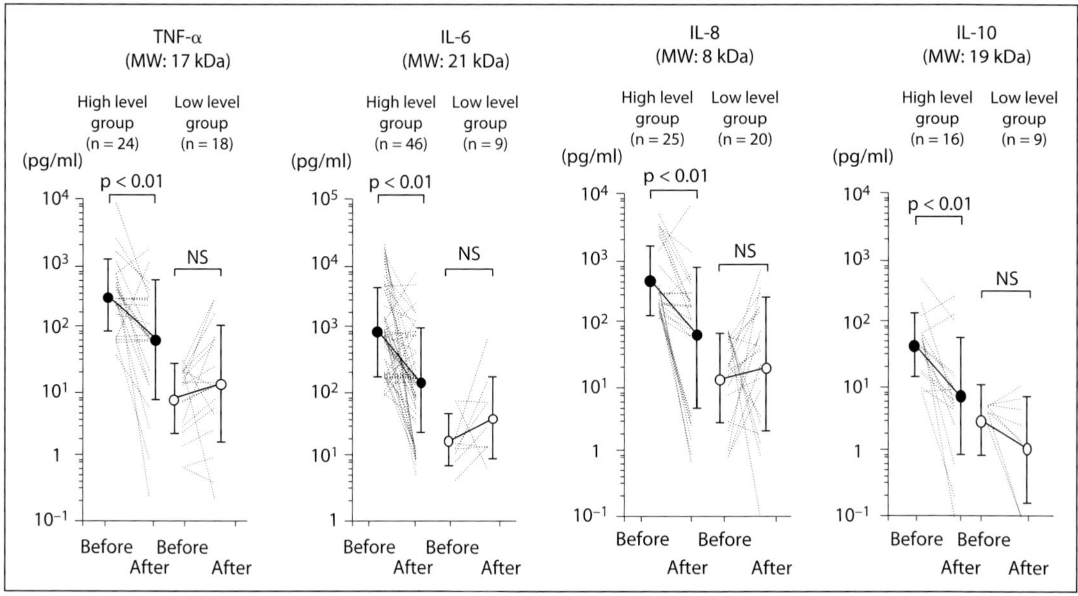

Fig. 1. Changes in blood levels of cytokines before and after 3 days' PMMA-CHDF treatment. Patients were divided into high and low level groups according to the initial levels of the cytokine of interest. Values are means ± SD; p values calculated by paired t test.

We previously examined the clinical efficacy of PMMA-CHDF in 43 patients with septic shock who were admitted to the ICU and underwent PMMA-CHDF within 24 h after onset of septic shock [15]. The baseline characteristics of these 43 patients at ICU admission were as follows: Acute Physiology and Chronic Health Evaluation (APACHE) II score 29.4 ± 8.4; Sequential Organ Failure Assessment (SOFA) score 13.7 ± 3.7; blood lactate level 72.2 ± 40.4 mg/dl; and blood IL-6 level 132,300 ± 243,700 pg/ml (means ± SD). A significant increase in blood pressure was observed during the first hour of PMMA-CHDF, with a subsequent further increase (fig. 2a). Recovery from shock was achieved in 39 patients (90.6%) and 34 patients survived for 28 days or longer (79.1%). Changes in blood IL-6 levels (as a biomarker of cytokine storm) and blood lactate levels (as a biomarker of dysoxia) were compared between survivors and non-survivors (fig. 2b). Regardless of outcomes, both blood IL-6 and lactate (which increased to extremely high levels) were significantly decreased after initiation of PMMA-CHDF. In the 34 survivors, both blood IL-6 and lactate decreased to values below the predetermined treatment goals (IL-6 <1,000 pg/ml; lactate <20 mg/dl) at 48 h after initiation of PMMA-CHDF. These findings suggested that cytokine removal by PMMA-CHDF led to an early recovery from dysoxia and a favorable outcome.

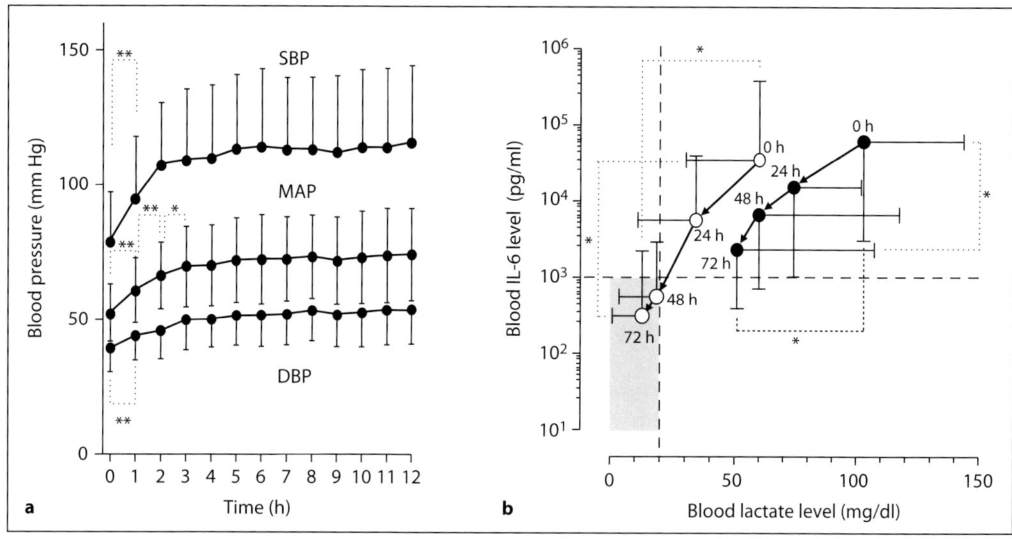

Fig. 2. Changes in various parameters after initiation of early PMMA-CHDF in 43 patients with septic shock. **a** Change in blood pressure (means ± SD). SBP = Systolic blood pressure; MBP = mean blood pressure; DBP = diastolic blood pressure. **b** Changes in blood levels of IL-6 and lactate (means ± SD). ○ = Survivors (n = 34), ● = non-survivors (n = 9). * $p < 0.05$, ** $p < 0.001$ (paired t test).

As another method of blood purification for the treatment of patients with septic shock, direct hemoperfusion with an endotoxin-adsorbing column containing immobilized polymyxin-B (PMX-DHP) is widely used, especially in Japan [16]. However, there are some questions regarding the therapeutic effect of PMX-DHP. Endotoxin is one of the pathogen-associated molecular patterns, and are not always the trigger of septic shock [6]. Therefore, removal of endotoxin alone does not lead to recovery from septic shock in all patients with septic shock. The therapeutic effect of PMX-DHP has sometimes also been questioned. The standard period of PMX-DHP therapy is only 2 h and since endotoxin is thought to be released from infectious focus or to be translocated from gut continuously, 2 h of treatment is too short for the improvement in septic shock. In some studies reporting the efficacy of PMX-DHP, another form of blood purification such as PMMA-CHDF was performed in addition to PMX-DHP [17]. Furthermore, there are no large multicenter trials on the efficacy of PMX-DHP. As a consequence, we have never applied PMX-DHP for the treatment of septic shock in our ICU [17]. The 43 patients mentioned above were all treated with PMMA-CHDF alone.

Table 1 shows a comparison of the outcome among our 43 patients treated with PMMA-CHDF and the outcomes in recently reported studies on patients with septic shock treated with HVHF [18–21] or PMX-DHP [22–25]. In most reports, the observed 28-day survival rate was higher than that predicted from

Table 1. Therapeutic effects of various blood purification treatments in septic shock

First author	Publication year	Patients, n	APACHE II	Predicted survival, %	Observed 28-day survival, %	Observed/ predicted survival ratio
HVHF						
Honore [18]	2000	20	31.5±4.2	26.7	55.0	2.06
Piccinni [19]	2006	40	27.2±2.8	39.5	55.0	1.39
Cornejo [20]	2006	20	26.1±3.1	43.1	60.0	1.39
Boussekey [21]	2008	9	30.3±4.1	29.7	66.7	2.25
PMX-DHP						
Vincent [22]	2005	17	16.7±5.9	73.8	70.6	0.97
Kojika [23]	2006	24	16.8±4.1	73.8	87.5	1.19
Nakamura [24]	2009	40	21.5±4.5	57.6	70.0	1.22
Cruz [25]	2009	34	21 (19–23)	61.1	67.7	1.11
PMMA-CHDF						
Our previous study [15]	2008	43	29.4±8.4	32.8	79.1	2.41

APACHE II scores presented as mean ± SD or median (range).

the APACHE II score, suggesting that the blood purification treatment tested was clinically effective. For our 43 septic patients treated with PMMA-CHDF [15], a high 28-day survival rate (79.1%) was observed despite a low predicted survival (32.8%), with an observed/predicted survival ratio of 2.41, the highest value among the studies listed in table 1. Due to differences in study design and patient characteristics among these studies, simple comparison of outcomes among these studies is difficult. Nevertheless, the data in table 1 suggest that the clinical efficacy of PMMA-CHDF is not inferior to that of any of the other blood purification methods with which it was compared.

Immunomodulatory Effects of Cytokine Removal by PMMA-CHDF on Immunocompetent Cells

In addition to early recovery from dysoxia, we have confirmed that cytokine removal by PMMA-CHDF in septic shock has other additional clinical benefits.

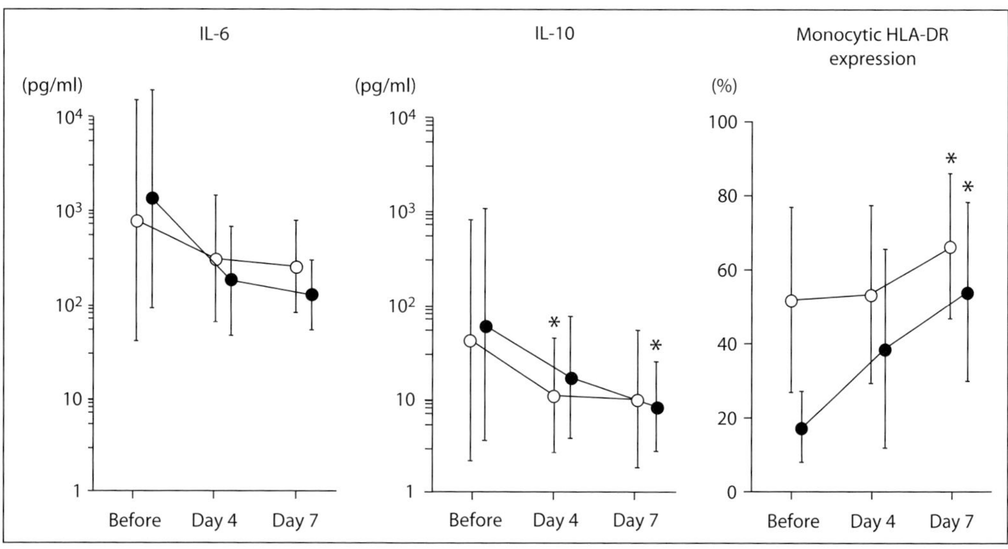

Fig. 3. Changes in blood IL-6 level, blood IL-10 level, and rate of monocytic HLA-DR expression during PMMA-CHDF in 16 patients with sepsis (means ± SD). ○ = All patients (n = 16); ● = patients with initial rate of monocytic HLA-DR expression <30% (n = 4). * $p < 0.05$ vs. before (paired t test).

Among them, we will discuss in this section the effects on cellular immunity. The delayed neutrophil apoptosis generally observed in patients with severe sepsis/septic shock is closely involved in the development of organ failure [2, 6]. We have observed that removal of pro-inflammatory cytokines by PMMA-CHDF in severe sepsis/septic shock resulted in recovery from delayed neutrophil apoptosis [26]. Furthermore, anti-inflammatory cytokines are thought to be involved in the development of immunoparalysis in the late stage of severe sepsis/septic shock [3]. A decrease in the monocytic HLA-DR expression rate has been suggested to be a useful diagnostic criterion for immunoparalysis [3]. Schefold et al. [27] have recently reported that DHP using a column conjugated with antibodies directed against IL-6, IL-10, and complement selectively removed these mediators from the blood stream, leading to recovery of the rate of monocytic HLA-DR expression. We demonstrated that PMMA-CHDF efficiently removed IL-10, an anti-inflammatory cytokine, from the blood stream (fig. 1) [11]. In our other study, it was also demonstrated that PMMA-CHDF treatment in sepsis significantly reduced blood levels of IL-10, with a concomitant increase in rate of monocytic HLA-DR expression, which had been depressed at baseline (fig. 3). These findings suggest an important future role for blood purification, an immunomodulatory treatment intended not only to correct altered cytokine profiles in the blood stream but also to restore the immunocompetent cells [28].

Our method of PMMA-CHDF thus appears to have the potential for use in immune cell modulatory therapy.

Enhancement of the Cytokine-Removing Capacity of PMMA-CHDF to Cope with Refractory Septic Shock

While PMMA-CHDF was a useful cytokine modulator for the treatment of patients with septic shock, as mentioned above, it was also demonstrated that some non-surviving patients with septic shock exhibited an abnormally high blood IL-6 level upon initiation of PMMA-CHDF [14, 29]. In patients with these abnormally high blood IL-6 levels, our standard PMMA-CHDF treatment sometimes failed to reduce it below the treatment goal (<1,000 pg/ml). The failure of standard PMMA-CHDF to treat these patients with extraordinarily severe cytokine storm may be due to insufficient cytokine modulation. Exacerbation of cytokine storm is thought to depend on 2 factors. One is the type of pathogen involved. Infection with a highly virulent pathogen inducing excess inflammatory response such as group A streptococci is known to induce severe cytokine storm [30]. The other is host genetic polymorphisms related to cytokine production. We have reported that extremely severe cytokine storm occurs in critically ill patients carrying high-risk alleles of genes related to cytokine production [29]. We are currently attempting to cope with exacerbation of cytokine storm (either caused by highly virulent pathogenic microorganisms or induced in patients with high-risk genotypes) by early initiation of double PMMA-CHDF, which involves 2 blood circuits for PMMA-CHDF per patient to enhance cytokine-removing capacity.

Conclusion

The clinical efficacy of PMMA-CHDF intended for cytokine removal and early improvements in shock and immunohomeostasis was examined in patients with severe sepsis/septic shock. Cytokine removal by PMMA-CHDF was effective in improvement of clinical condition and outcome in severe sepsis/septic shock. Since PMMA-CHDF is gently, continuously and easily performed under operating conditions similar to those employed for routine renal support, it is expected to be useful as a means of blood purification treatment for critically ill patients. We are planning to conduct a randomized trial to demonstrate the therapeutic effects of PMMA-CHDF in severe sepsis/septic shock and then promote its wide clinical application.

References

1 Martin GS, Mannino DM, Eaton S, Moss M: The epidemiology of sepsis in the United States from 1979 through 2000. N Engl J Med 2003;348:1546–1554.

2 Hotchkiss RS, Karl IE: The pathophysiology and treatment of sepsis. N Engl J Med 2003; 348:138–150.

3 Volk HD, Reinke P, Döcke WD: Clinical aspects: from systemic inflammation to 'immunoparalysis'. Chem Immunol 2000;74: 162–177.

4 Ronco C, Tetta C, Mariano F, Wratten ML, Bonello M, Bordoni V, Cardona X, Inguaggiato P, Pilotto L, d'Intini V, Bellomo R: Interpreting the mechanism of continuous renal replacement therapy in sepsis: the peak concentration hypothesis. Artif Organs 2003;27:792–801.

5 Webster NR, Galley HF: Immunomodulation in the critically ill. Br J Anaesth 2009;103:70–81.

6 Cinel I, Opal SM: Molecular biology of inflammation and sepsis: a primer. Crit Care Med 2009;37:291–304.

7 De Vriese AS, Colardyn FA, Philippé JJ, Vanholder RC, De Sutter JH, Lameire NH: Cytokine removal during continuous hemofiltration in septic patients. J Am Soc Nephrol 1999;10:846–853.

8 Sieberth HG, Kierdorf HP: Is cytokine removal by continuous hemofiltration feasible? Kidney Int Suppl 1999;72:S79–S83.

9 Honore PM, Joannes-Boyau O, Boer W, Collin V: High-volume hemofiltration in sepsis and SIRS: current concepts and future prospects. Blood Purif 2009;28:1–11.

10 Haase M, Bellomo R, Morgera S, Baldwin I, Boyce N: High cut-off point membranes in septic acute renal failure: systematic review. Int J Artif Organs 2007;30:1031–1041.

11 Matsuda K, Hirasawa H, Oda S, Shiga H, Nakanishi K: Current topics on cytokine removal technologies. Ther Apher 2001;5: 306–314.

12 Oda S, Hirasawa H, Shiga H, Nakanishi K, Matsuda K, Nakamura M: Continuous hemofiltration/hemodiafiltration in critical care. Ther Apher 2002;6:193–198.

13 Hirasawa H, Oda S, Matsuda K: Continuous hemodiafiltration with cytokine-adsorbing hemofilter in the treatment of severe sepsis and septic shock. Contrib Nephrol 2007;156: 365–370

14 Oda S, Hirasawa H, Shiga H, Nakanishi K, Matsuda K, Nakamua M: Sequential measurement of IL-6 blood levels in patients with systemic inflammatory response syndrome (SIRS)/sepsis. Cytokine 2005;29:169–175.

15 Nakada TA, Oda S, Matsuda K, Sadahiro T, Nakamura M, Abe R, Hirasawa H: Continuous hemodiafiltration with PMMA hemofilter in the treatment of patients with septic shock. Mol Med 2008;14:257–263.

16 Shimizu T, Endo Y, Tsuchihashi H, Akabori H, Yamamoto H, Tani T: Endotoxin apheresis for sepsis. Transfus Apher Sci 2006;35:271–282.

17 Hirasawa H, Oda S, Shiga H, Matsuda K, Nakanishi K: Endotoxin adsorption or hemodiafiltration in the treatment of multiple organ failure. Curr Opin Crit Care 2000; 6:421–425.

18 Honore PM, Jamez J, Wauthier M, Lee PA, Dugernier T, Pirenne B, Hanique G, Matson JR : Prospective evaluation of short-term, high-volume isovolemic hemofiltration on the course and outcome in patients with intractable circulatory failure resulting from septic shock. Crit Care Med 2000;28:3581–3587.

19 Piccinni P, Dan M, Carraro R, Carraro R, Lieta E, Marafon S, Zamperetti N, Brendolan A, D'Intini V, Tetta C, Bellomo R, Ronco C: Early isovolemic haemofiltration in oligouric patients with septic shock. Intensive Care Med 2006;32:80–86.

20 Cornejo R, Downey P, Castro R, Romero C, Regueira T, Vega J, Castillo L, Andresen M, Dougnac A, Bugedo G, Hernandez G: High-volume hemofiltration as salvage therapy in severe hyperdynamic septic shock. Intensive Care Med 2006;32:713–722.

21 Bousskey N, Chiche A, Faure K, Devos P, Guery B, d'Escrivan T, Georges H, Leroy O: A pilot study comparing high and low volume hemofiltration on vasopressor use in septic shock. Intensive Care Med 2008;34: 1646–1653.

22 Vincent JL, Laterre PF, Cohen J, Burchardi H, Bruining H, Lerma FA, Wittebole X, De Backer D, Brett S, Marzo D, Nakamura H, John S: A pilot-controlled study of a polymyxin B-immobilized hemoperfusion cartridge in patients with severe sepsis secondary to intra-abdominal infection. Shock 2005;23:400–405.
23 Kojika M, Sato N, Yaegashi Y, Suzuki Y, Suzuki K, Nakae H, Endo S: Endotoxin adsorption therapy for septic shock using polymyxin B-immobilized fibers (PMX): evaluation by high-sensitivity endotoxin assay and measurement of the cytokine production capacity. Ther Apher Dial 2006;10: 12–18.
24 Nakamura T, Sugaya T, Koide H: Urinary liver-type fatty acid-binding protein in septic shock: effect of polymyxin B-immobilized fiber hemoperfusion. Shock 2009;31:454–459.
25 Cruz DN, Antonelli M, Fumagalli R, Foltran F, Brienza N, Donati A, Malcangi V, Petrini F, Volta G, Bobbio Pallavicini FM, Rottoli F, Giunta F, Ronco C: Early use of polymyxin B hemoperfusion in abdominal septic shock: the EUPHAS randomized controlled trial. JAMA 2009;301:2445–2452.
26 Hirano T, Hirasawa H, Oda S, Shiga H, Nakanishi K, Matsuda K, Nakamura M, Asai T, Kitamura N: Modulation of polymorphonuclear leukocyte apoptosis in the critically ill by removal of cytokines with continuous hemodiafiltration. Blood Purif 2004;22:188–197.
27 Schefold JC, Haehling S, Corsepius M, Pohle C, Kruschke P, Zuckermann H, Volk HD, Reinke P: A novel selective extracorporeal intervention in sepsis: immunoadsorption of endotoxin, interleukin 6, and complement-activating product 5A. Shock 2007;28:418–425.
28 Schefold JC, Hasper D, Jorres A: Organ crosstalk in critically ill patients: hemofiltration and immunomodulation in sepsis. Blood Purif 2009;28:116–123.
29 Watanabe E, Hirasawa H, Oda S, Matsuda K, Hatano M, Tokuhisa T: Extremely high interleukin-6 blood levels and outcome in the critically ill are associated with tumor necrosis factor- and interleukin-1-related gene polymorphisms. Crit Care Med 2005;33:89–97.
30 Smith A, Lamagni TL, Oliver I, Efstratiou A, George RC, Stuart JM: Invasive group A streptococcal disease: should close contacts routinely receive antibiotic prophylaxis? Lancet Infect Dis 2005;5:494–500.

Masataka Nakamura
Department of Emergency and Critical Care Medicine, Chiba University Graduate School of Medicine
Inohana 1–8–1 Chuoku
260-8677 Chiba City, Chiba (Japan)
Tel. +81 43 226 2341, Fax +81 43 226 2371, E-Mail masatakan@faculty.chiba-u.jp

Suzuki H, Hirasawa H (eds): Acute Blood Purification.
Contrib Nephrol. Basel, Karger, 2010, vol 166, pp 83–92

Efficacy of Continuous Hemodiafiltration with a Cytokine-Adsorbing Hemofilter in the Treatment of Acute Respiratory Distress Syndrome

Kenichi Matsuda[a] · Takeshi Moriguchi[a] · Shigeto Oda[b] · Hiroyuki Hirasawa[b]

[a]Department of Emergency and Critical Care Medicine, University of Yamanashi School of Medicine, Yamanashi, and [b]Department of Emergency and Critical Care Medicine, Graduate School of Medicine, Chiba University, Chiba, Japan

Abstract

Background/Aims: In the pathophysiology of acute respiratory distress syndrome (ARDS), the increase in capillary and alveolar permeability caused by various humoral mediators and resultant pulmonary interstitial edema play major roles. In this study, the efficacy of continuous hemodiafiltration using a cytokine-adsorbing hemofilter with a membrane made of polymethylmethacrylate (PMMA-CHDF) in the treatment of ARDS patients was investigated. **Materials and Methods:** Fifty-one patients with a diagnosis of ARDS complicated by renal failure and without prior steroid therapy were enrolled in this study. Changes in respiratory index (RI), positive end-expiratory pressure, central venous pressure (CVP) and blood levels of TNFα, IL-6 and IL-8 before/after blood purification for 3 days as well as the cumulative water balance during the 3-day treatment and 28-day cumulative survival rate were compared between 2 patient groups. One group underwent PMMA-CHDF and the other intermittent hemodialysis (IHD) without water removal for elimination of metabolites and continuous hemofiltration (CHF) for fluid management. **Results:** Blood purification for 3 days significantly decreased blood levels of cytokines and successfully removed water without changing CVP in the PMMA-CHDF group, but not in the IHD+CHF group. Significant correlations between changes in blood levels of cytokines (IL-6 and IL-8) and changes in RI were demonstrated in the PMMA-CHDF group. The 28-day cumulative survival rate in the PMMA-CHDF group (68.8%) was significantly higher than that in the IHD+CHF group (36.8%). **Conclusions:** Cytokine removal therapy with PMMA-CHDF is expected to be useful as a new therapeutic modality in ARDS patients for non-renal indications.

The pathophysiology and treatment of acute respiratory distress syndrome (ARDS) have been extensively investigated. Recently, it has become generally accepted that the increases in capillary and alveolar permeability caused by various humoral mediators and resultant pulmonary interstitial edema play major roles in the pathogenesis of ARDS [1, 2]. The improved understanding of the pathophysiology of ARDS has provoked numerous novel therapeutic strategies [3–5]. Although many kinds of new approaches have been tried in the treatment of ARDS patients, the mortality rate of ARDS patients remains unacceptably high [6, 7].

We have previously reported the usefulness of continuous hemodiafiltration using a polymethylmethacrylate membrane hemofilter (PMMA-CHDF), which removes cytokines from the bloodstream by adsorption rather than diffusion and filtration [8], in the treatment of hypercytokinemia such as severe sepsis, septic shock and multiple organ failure [9–11].

In the present study, we investigated the efficacy of cytokine removal therapy with PMMA-CHDF in the treatment of ARDS, based on our previous report on the clinical usefulness of cytokine removal therapy with PMMA-CHDF in ARDS [12].

Materials and Methods

Subjects or Patients

The treatment group of the present study consisted of 32 patients admitted to the ICU of Chiba University Hospital between January 2002 and December 2006 with a diagnosis of ARDS complicated by renal failure and without prior steroid therapy. PMMA-CHDF was performed in all of them for renal replacement and cytokine removal (PMMA-CHDF group).

The control group of the present study consisted of 19 patients admitted to the ICU between 1995 and 2001 with a diagnosis of ARDS complicated by renal failure and without prior steroid therapy. Intermittent hemodialysis (IHD) without water removal for elimination of metabolites (such as blood urea nitrogen, BUN, and creatinine) was performed in all of them with concomitant continuous hemofiltration (CHF) for fluid management (IHD+CHF group).

ARDS was diagnosed according to the criteria provided in the report of the American-European consensus conference on ARDS [1]. Renal failure was diagnosed when all of the following criteria were met: non-responsiveness to water loading and diuretics, BUN >50 mg/dl and creatinine >3 mg/dl.

Outcome Measures

First, Acute Physiology and Chronic Health Evaluation (APACHE) II score, respiratory index (RI), positive end-expiratory pressure (PEEP), central venous pressure (CVP) and blood levels of TNF-α, IL-6 and IL-8 at baseline were compared between the PMMA-CHDF and IHD+CHF groups.

Second, changes in RI, PEEP, CVP and blood levels of TNF-α, IL-6 and IL-8 after 3 days as well as cumulative water balance (CWB) were compared between the PMMA-

CHDF and IHD+CHF groups. The correlations between changes in blood levels of TNF-α, IL-6 and IL-8 (ΔTNF-α, ΔIL-6, ΔIL-8) and change in RI (ΔRI) were also examined.

Finally, the cumulative 28-day survival rates of the PMMA-CHDF and IHD+CHF groups were compared.

Measurements and Assay Methods

Blood samples for blood gas analysis and determination of cytokine level were collected via a catheter inserted into the radial artery. RI was calculated from the results of blood gas analysis using the following equation: RI = $AaDO_2/PaO_2$.

CVP was measured by a pressure transducer (Baxter International, Deerfield, Ill., USA) and a central monitoring system (CNS-8300, Nihon Koden, Tokyo, Japan) with a catheter inserted into the ascending vena cava through the jugular, subclavian or femoral veins.

In the PMMA-CHDF group, blood levels of IL-6 were measured using freshly isolated plasma with a rapid IL-6 assay system based on the chemiluminescent enzyme immunoassay (Human IL-6 CLEIA cartridge and Lumipulse, Fujirebio, Tokyo, Japan) [8], while blood levels of TNF-α and IL-8 were measured using plasma stored at –80°C after centrifugal separation by ELISA (TNFα EASIA, BioSource Europe, Nivelles, Belgium; Human IL-8 ELISA kit, Fujirebio). In the IHD+CHF group, in contrast, the blood levels of all 3 cytokines were determined using plasma stored at –80°C after centrifugal separation by ELISA (TNF-α EASIA and IL-6 EASIA).

Procedures for PMMA-CHDF

Vascular access was obtained via a flexible double-lumen catheter (Arrow International, Reading, Pa., USA) placed in the internal jugular, femoral or subclavian vein. A bedside console specially designed for CHDF (JUN-500, Ube Medical, Tokyo, Japan) was used to monitor operation of the hemodiafiltration system. A PMMA membrane hemofilter capable of removing humoral mediators at high efficiency (Hemofeel CH1.0L, Toray Medical, Tokyo, Japan; membrane surface area, 1.0 m^2) was used [8–11]. Nafamostat mesilate (Futhan, Torii Pharmaceutical, Tokyo, Japan) was used as anticoagulant. It is a synthetic proteinase inhibitor with an anticoagulant property, and we have reported that this protease inhibitor reduces hemorrhagic complications during CHDF [13].

An electrolyte solution appropriate for the patient's condition was used as replacement fluid and supplied in post-dilution mode. A sterile bicarbonate replacement fluid (Sublood-B, Fuso Pharmaceutical Industries, Osaka, Japan) was used as dialysate. Basic operating conditions for CHDF were as follows: blood flow rate 60–120 ml/min, filtration rate 300–500 ml/h and dialysate flow rate 500–2,000 ml/h. These conditions were modified as necessary, depending on the patient's condition.

Procedures for IHD+CHF

IHD without water removal was performed for elimination of metabolites such as BUN and creatinine. Vascular access was obtained via a flexible double-lumen catheter placed in the internal jugular, femoral or subclavian veins. A personal bedside console for hemodialysis was used to monitor operation of the hemodialysis system. IHD was performed for 3–4 h each time and 3–4 times a week, depending on the patient's clinical condition. A dialyzer with membrane made of regenerated cellulose, cellulose diacetate or ethylene vinyl alcohol copolymer was used. The membrane surface area of the dialyzer was 1.0–2.0 m^2, and was varied depending on the patient's body size and clinical condition. Nafamostat mesilate, low-molecular-weight heparin or heparin was used as anticoagulant. A bicarbonate dialysate was used as dialysate. Basic operating conditions for IHD were a blood flow rate of 100–150 ml/min (to minimize effects on circulatory dynamics) and a fixed dialysate flow rate of 500 ml/h.

CHF for management of body fluid was continuously performed, with additional IHD. Vascular access was obtained in the same fashion as for IHD. A bedside console specially designed for CHF (CHF-1, Ube Medical, Tokyo, Japan) was used to monitor operation of the hemofiltration system. A hemofilter with membrane made of polyacrylonitrile or polysulfone was used. The membrane surface area of the hemofilter was 0.3–0.6 m^2, and was varied depending on the patient's body size and clinical condition. Nafamostat mesilate or low-molecular-weight heparin was used as anticoagulant. An electrolyte solution appropriate for the patient's condition was used as replacement fluid and supplied in post-dilution mode. Basic operating conditions for the hemofiltration system were a blood flow rate of 60–100 ml/min and a filtration rate of 50–100 ml/h depending on the patient's clinical condition.

Mechanical Ventilation

Mechanical ventilation in patients with ARDS was provided in a similar fashion in the 2 groups. A mechanical ventilator was operated under the following conditions: ventilation mode, volume control, or pressure control; tidal volume, 8–10 ml/kg. Lung-protective ventilation with permissive hypercapnia [1, 3, 5], which is currently recommended for the treatment of ARDS, was performed in neither the PMMA-CHDF nor the IHD+CHF group during the present study.

Statistics

Values are presented as means ± SD. Baseline data were compared between the PMMA-CHDF and IHD+CHF groups by Student's unpaired t test and the Wilcoxon signed-rank test. The statistical significance of changes in values before and after blood purification for 3 days observed within each group was tested by Student's paired t test. In addition, pre- to post-treatment changes were compared between the 2 groups by repeated-measures ANOVA and Student's unpaired t test. The 28-day cumulative survival rates were calculated for the 2 groups by the Kaplan-Meier method and statistically compared with the Mantel-Cox test. Values of $p < 0.05$ were considered significant. The StatView 5.0 Software Package for Mac (SAS Institute, Cary, N.C., USA) was used for all statistical analyses.

Results

The demographic and physiologic variables of the ARDS patients with renal failure in the PMMA-CHDF and IHD+CHF groups before the treatment are presented in table 1. While RI and blood level of IL-6 were significantly higher in the PMMA-CHDF group, no significant differences in other parameters were noted between the 2 groups.

Table 1. Demographic and physiologic variables of ARDS patients with acute renal failure before the treatment (means ± SD)

	PMMA-CHDF group (n = 32)	IHD+CHF group (n = 19)	p value
Age, years	61.5±15.0	55.8±15.1	n.s.
Background			
Sepsis	26	13	
Peritonitis	11	7	
Pneumonia	7	3	
Other	8	3	
Other	6	6	
APACHE II score	24.6±8.6	26.9±8.5	n.s.
RI	5.5±2.4	4.0±2.1	<0.05
PEEP, cm H_2O	4.0±2.8	3.3±3.1	n.s.
CVP, cm H_2O	7.3±4.4	7.5±3.3	n.s.
TNFα(log), pg/ml	1.90±0.58 (79)	1.78±0.35 (60.1)	n.s.
IL-6(log), pg/ml	3.64±1.00 (4,360)	2.32±0.57 (209)	<0.05
IL-8(log), pg/ml	2.29±0.61 (195)	1.90±0.68 (79.3)	n.s.

Table 2. Changes in physiological variables, blood level of cytokines and CWB after 3 days of treatment.

	PMMA-CHDF group		IHD+CHF	
	before	after	before	after
RI	5.5±2.4[b]	2.8±1.5[a, c]	4.0±2.1	2.2±2.5[a]
PEEP, cm H_2O	4.0±2.8	4.3±2.3	3.3±3.1	4.0±3.3
CVP, cm H_2O	7.3±4.4	8.5±4.0	7.5±3.3	8.1±5.1
TNFα (log), pg/ml	1.90±0.58 (79)	1.68±0.39 (48)	1.78±0.35 (60)	1.96±0.25 (91)
IL-6 (log), pg/ml	3.64±1.00[b] (4,360)	2.59±0.59[a] (390)	2.32±0.57 (210)	2.50±0.70 (320)
IL-8 (log), pg/ml	2.29±0.61 (195)	1.73±0.43[a] (54)	1.90±0.68 (79)	1.92±0.71 (83)
CWB (3 days), ml	–975[b]		305	

[a] $p < 0.05$ pre vs. post; [b] $p < 0.05$ PMMA vs. IHD+CHF, [c] $p < 0.05$ PMMA vs. IHD+CHF in the degree of changes following the 3-day treatment.

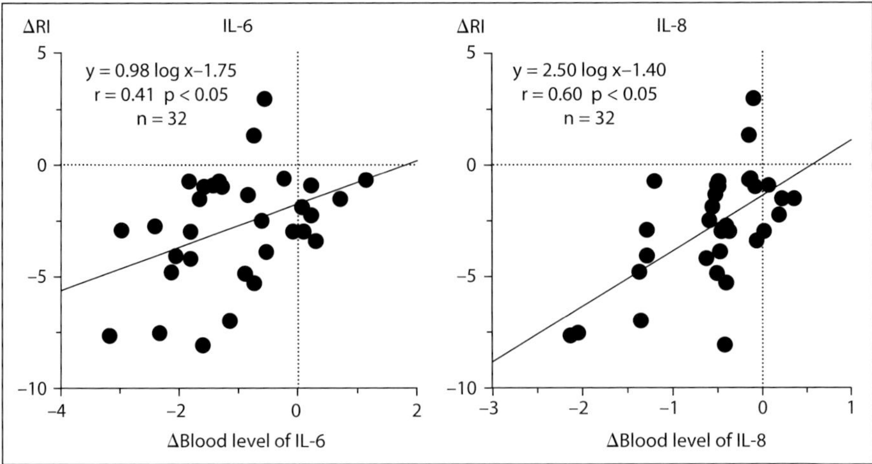

Fig. 1. Correlations between changes in blood levels of cytokines on a logarithmic scale (before/after treatment) and ΔRI after 3 days' treatment in the PMMA-CHDF group. There were significant and positive correlations between the changes in blood levels of IL-6 and ΔRI and between changes in blood levels of IL-8 and ΔRI.

Table 2 shows changes in physiologic variables, blood level of cytokines and CWB following 3 days of treatment. Three-day treatment significantly improved RI in both patient groups without changing PEEP and CVP. Although baseline RI was significantly higher in the PMMA-CHDF group, no significant difference in post-treatment RI was noted between the groups. The degrees of improvement in RI achieved by blood purification for 3 days in the 2 groups were therefore compared by repeated-measures ANOVA, which demonstrated that improvement in RI was significantly greater in the PMMA-CHDF group than in the IHD+CHF group. Significant decreases in blood levels of cytokines were observed for the 3-day treatment period in the PMMA-CHDF group, but not in the IHD-CHF group. The CWB for the 3-day treatment period was significantly lower in the PMMA-CHDF group (–975 ml) than in the IHD+CHF group (+305 ml).

Correlations between changes in blood levels of cytokines and ΔRI during blood purification for 3 days were examined separately for each group. In the PMMA-CHDF group, significant positive correlations were observed between ΔIL-6 and ΔRI as well as between ΔIL-8 and ΔRI (fig. 1). In the IHD+CHF group, in contrast, no significant correlation was observed between the change in blood level of any cytokine and ΔRI (data not shown).

A Kaplan-Meier plot depicting survivors of ARDS patients with renal failure during the 28-day period are demonstrated in figure 2. The 28-day survival rate in the PMMA-CHDF group (68.8%) was significantly higher than that in the IHD+CHF group (36.8%).

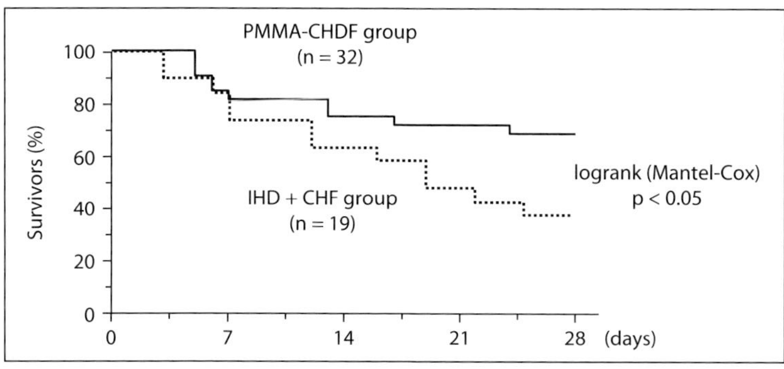

Fig. 2. Kaplan-Meier plot depicting the survivors of ARDS patients with renal failure during the 28-day period. The 28-day survival in the PMMA-CHDF group was 68.8%, which was significantly better than 36.8% in the IHD+CHF group.

Discussion

Types of continuous renal replacement therapy such as CHF and CHDF have been found to stabilize hemodynamics in patients with sepsis and other inflammatory syndromes, as well as to improve pulmonary oxygenation in patients with respiratory failure [14, 15]. Since these clinical effects were demonstrated even when continuous renal replacement therapy was performed without water removal, several investigators have attributed these effects not to removal of water but to removal of causative humoral mediators of hyperpermeability such as cytokines from the bloodstream [16, 17]. Although clinical efficacy of cytokine removal therapy with blood purification has been demonstrated in individual patients with hypercytokinemia, its clinical usefulness has been difficult to demonstrate in the setting of a randomized controlled trial and thus remains to be established [18, 19]. Therefore, cytokine removal therapy with blood purification has attracted no attention in the treatment of ARDS patients [1, 3, 5].

Based on our previous report on the clinical usefulness of cytokine removal therapy with PMMA-CHDF in ARDS [12], the present study investigated in further detail the clinical usefulness of this method of treatment for ARDS.

Cytokine removal therapy with PMMA-CHDF performed in ARDS patients with renal failure significantly reduced blood levels of cytokines, while cytokine removal with IHD+CHF failed (table 2). This finding confirmed our previous reports that PMMA-CHDF reduced blood levels of cytokines in patients with hypercytokinemia [8–11].

CWB during blood purification for 3 days was also examined in the present study (table 2). In the PMMA-CHDF group, 975 ml water was removed during the treatment without changing CVP. In the IHD+CHF group, in contrast,

CWB during the 3-day IHD+CHF treatment was +305 ml. These observations demonstrated that water was removed from the pulmonary interstitial tissue without affecting hemodynamics in the PMMA-CHDF group. The pathogenesis of ARDS is known to involve injury and increased permeability of alveolar epithelial cells and pulmonary capillary endothelial cells due to excess production of humoral mediators by immunocompetent cells, leading to subsequent interstitial edema [1, 2]. Removal of humoral mediators causing ARDS by PMMA-CHDF should thus improve pulmonary capillary permeability, leading to refilling of water from the pulmonary interstitial tissue back into the bloodstream. Furthermore, water refilling into the bloodstream could be removed by PMMA-CHDF, yielding a negative CWB value without concomitant decrease in CVP. Efficient removal of cytokines and further removal of water from the pulmonary interstitial tissue to reduce interstitial edema together explain the greater improvement of pulmonary oxygenation observed in the PMMA-CHDF group compared with that in the IHD+CHF group (table 2). Furthermore, decreases in blood levels of IL-6 and IL-8 were significantly correlated with improvement in pulmonary oxygenation in the PMMA-CHDF group (fig. 1), suggesting the importance of cytokine removal for improving pulmonary oxygenation in patients with ARDS.

Although attempts to treat ARDS with various modalities have been reported, clinical usefulness has been clearly established only for lung-protective ventilation with permissive hypercapnia [1–5]. Since the present study suggested the possibility that cytokine removal in ARDS patients not only improved pulmonary oxygenation but also increased survival rate (fig. 2), cytokine removal therapy with PMMA-CHDF may prove useful as a new treatment modality for ARDS.

The present study has some limitations. It was a retrospective study including patients treated prior to the introduction of PMMA-CHDF as historical controls. In the ICU of our hospital, PMMA-CHDF was introduced in October 1989 as renal replacement therapy for patients with acute renal failure. Since then, we have experienced many cases in which PMMA-CHDF increased urine volume and stabilized hemodynamics immediately after its initiation, which led us to adopt it as our first-line blood purification therapy for treatment of acute renal failure [20]. As a consequence, the decision not to perform PMMA-CHDF in patients in need of renal replacement became ethically impossible in the ICU. This is the reason for our use of a retrospective study design involving comparison with historical controls. Although the cases in the IHD+CHF group were experienced earlier (1995–2001) than those in the PMMA-CHDF group (2002–2006), lung-protective ventilation with permissive hypercapnia, which is the only treatment modality for ARDS with demonstrated clinical efficacy, was not performed in either of the 2 groups. We therefore believe that the differences in treatment for ARDS between these 2 groups were minor, except for the blood purification technologies employed.

Based on the findings of this study, we are planning to investigate the clinical usefulness of cytokine removal therapy with PMMA-CHDF in ARDS patients without renal failure in a randomized controlled trial.

Conclusions

The efficacy of CHDF using a cytokine-adsorbing hemofilter with a membrane made of PMMA in the treatment of ARDS patients was investigated. The findings obtained suggested that cytokine removal and concomitant water removal from the pulmonary interstitial tissue by PMMA-CHDF might improve pulmonary oxygenation and consequently increase the survival rate of ARDS. Cytokine removal therapy with PMMA-CHDF is expected to be useful as a new treatment modality for ARDS, and deserves a prospective randomized controlled trial to establish the efficacy of CHDF with cytokine-adsorbing hemofilter in the treatment of ARDS without renal failure.

References

1 Wheeler AP, Bernard GR: Acute lung injury and the acute respiratory distress syndrome: a clinical review. Lancet 2007;369:1553–1565.

2 Maniatis NA, Kotanidou A, Catravas JD, Orfanos SE: Endothelial pathomechanisms in acute lung injury. Vascul Pharmacol 2008; 49:119–133.

3 Hemmila MR, Napolitano LM: Severe respiratory failure: advanced treatment options. Crit Care Med 2006;34(suppl):S278–S290.

4 Raghavendran K, Pryhuber GS, Chess PR, Davidson BA, Knight PR, Notter RH: Pharmacotherapy of acute lung injury and acute respiratory distress syndrome. Curr Med Chem 2008;15:1911–1924.

5 Antonelli M, Azoulay E, Bonten M, Chastre J, Citerio G, Conti G, De Backer D, Lemaire F, Gerlach H, Groeneveld J, Hedenstierna G, Macrae D, Mancebo J, Maggiore SM, Mebazaa A, Metnitz P, Pugin J, Wernerman J, Zhang H: Year in review in Intensive Care Medicine, 2008. II. Experimental, acute respiratory failure and ARDS, mechanical ventilation and endotracheal intubation. Intensive Care Med 2009;35:215–231.

6 Zambon M, Vincent JL: Mortality rates for patients with acute lung injury/ARDS have decreased over time. Chest 2008;133:1120–1127.

7 Phua J, Badia JR, Adhikari NK, Friedrich JO, Fowler RA, Singh JM, Scales DC, Stather DR, Li A, Jones A, Gattas DJ, Hallett D, Tomlinson G, Stewart TE, Ferguson ND: Has mortality from acute respiratory distress syndrome decreased over time? A systematic review. Am J Respir Crit Care Med 2009;179: 220–227.

8 Matsuda K, Hirasawa H, Oda S, Shiga H, Nakanishi K: Current topics on cytokine removal technologies. Ther Apher 2001;5: 306–314.

9 Nakada T, Hirasawa H, Oda S, Shiga H, Matsuda K: Blood purification for hypercytokinemia. Transfus Apher Sci 2006;35:253–264.

10 Hirasawa H, Oda S, Matsuda K: Continuous hemodiafiltration with cytokine-adsorbing hemofilter in the treatment of severe sepsis and septic shock. Contrib Nephrol 2007;156: 365–370.

11 Nakada T, Oda S, Matsuda K, Sadahiro T, Nakamura M, Abe R, Hirasawa H: Continuous hemodiafiltration with PMMA hemofilter in the treatment of patients with septic shock. Mol Med 2008;14:257–263.

12 Hirasawa H, Sugai T, Oda S, Shiga H, Matsuda K, Ueno H, Sadahiro T, Hikita S: Continuous hemodiafiltration (CHDF) removes cytokines and improves respiratory index (RI) and oxygen metabolism in patients with acute respiratory distress syndrome (ARDS). Crit Care Med 1998;26 (suppl):A120.
13 Ohtake Y, Hirasawa H, Sugai T, Oda S, Shiga H, Matsuda K, Kitamura N: Nafamostat mesylate as anticoagulant in continuous hemofiltration and continuous hemodiafiltration. Contrib Nephrol 1991;93:215–217.
14 Schetz M: Non-renal indications for continuous renal replacement therapy. Kidney Int 1999;72:S88–S94.
15 Zimmerman JL: Respiratory failure. Blood Purif 2002;20:235–238.
16 DiCarlo JV, Alexander SR, Agarwal R, Schiffman JD: Continuous veno-venous hemofiltration may improve survival from acute respiratory distress syndrome after bone marrow transplantation or chemotherapy. J Pediatr Hematol Oncol 2003;25:801–805.
17 Piccinni P, Dan M, Barbacini S, Carraro R, Lieta E, Marafon S, Zamperetti N, Brendolan A, D'Intini V, Tetta C, Bellomo R, Ronco C: Early isovolaemic haemofiltration in oliguric patients with septic shock. Intensive Care Med 2006;32:80–86.
18 Ronco C: Renal replacement therapy for acute kidney injury: let's follow the evidence. Int J Artif Organs 2007;30:89–94.
19 House AA, Ronco C: Extracorporeal blood purification in sepsis and sepsis-related acute kidney injury. Blood Purif 2008;26:30–35.
20 Hirayama Y, Hirasawa H, Oda S, Shiga H, Nakanishi K, Matsuda K, Nakamura M, Hirano T, Moriguchi T, Watanabe E, Nitta M, Abe R, Nakada T: The change in renal replacement therapy on acute renal failure in a general ICU in a university hospital and its clinical efficacy: a Japanese experience. Ther Apher Dial 2003;7:475–482.

Kenichi Matsuda, MD
Department of Emergency and Critical Care Medicine, University of Yamanashi School of Medicine
1110 Shimokato Chuo
Yamanashi, 409-3898 (Japan)
Tel. +81 55 273 9812, Fax +81 55 273 6716, E-Mail matsudak@yamanashi.ac.jp

Suzuki H, Hirasawa H (eds): Acute Blood Purification.
Contrib Nephrol. Basel, Karger, 2010, vol 166, pp 93–99

Blood Purification for Intoxication

Hajime Nakae

Department of Emergency and Critical Care Medicine, Akita University Graduate School of Medicine, Akita, Japan

Abstract

Blood purification is administered in cases of acute intoxication when the substance causing the intoxication is to be eliminated or when the substance leads to a case of organ dysfunction, such as in renal or hepatic failure. The causative substances cover a wide range, from medical drugs or agrichemicals to natural poisons (such as poisonous mushrooms). In removing these substances, gastric lavage, activated carbon administration, laxative administration or enema cleaning are the preferred methods, and blood purification is not routinely conducted. However, when the causative substance is unknown or when there are several causative substances, it is not easy to immediately grasp the disposition of the patient and so judge whether or not blood purification should be performed. In such cases, blood purification must be conducted in a timely manner and in accordance with the crisis management principle of 'prepare for the worst'. In general, substances whose molecular weight is within the removal spectrum, having a small distribution volume and a low protein-binding rate, are easier to remove. For substances with high protein-binding rates, albumin dialysis (MARS® and Prometheus®) is performed in order to remove albumin-binding substances. Since MARS and Prometheus® have not been introduced in Japan, plasma diafiltration, employing selective plasma filtration with dialysis, is a practical alternative.

As a general rule of treatment for cases of acute intoxication, information concerning the toxic substance that is the cause of the intoxication must be obtained immediately, and any unabsorbed substance must be removed, while that which has already been absorbed into the body must be discharged. The purpose of blood purification is to facilitate the discharge of this absorbed substance. This article describes the suitability of blood purification in these cases and responses to protein-binding substances.

Indications for Blood Purification

Blood purification is conducted in order to: (1) rapidly remove toxic substances in the blood; (2) remove toxic and other accumulated metabolites due to organ dysfunction resulting from intoxication; (3) control the balance between water and electrolytes; (4) replenish useful substances. There are various methods for purifying blood, which should be carefully selected by taking into consideration the volume of distribution (V_D) of the intoxicating substance and the protein-binding rate before implementation. The volume of distribution indicates the relation between the intake amount and the plasma concentration of the intoxicating substance. If the intoxicating substance is evenly distributed in the body fluid, the value $V_D = 0.6$ l/kg, is obtained. A value of V_D exceeding 0.6 l/kg means that the transition rate of the intoxicating substance is high because of its lipid solubility. In other words, the larger the volume of distribution is, the less the quantity of the intoxicating substance in plasma becomes. Simply put, the effectiveness of blood purification is lowered. Generally, the capacitance value $V_D <1$ l/kg is the range where blood purification is suitable.

The intoxicating substances in blood are divided into 2 types: a bound type that is linked to plasma protein and a free type which is not linked. The bound substance can not permeate cell membranes and its toxicity can not be drawn forth. Only the free substance permeates the cell membrane and allows the toxicity to be drawn out. As blood purification is a method of removing this free-type substance, the efficiency of removal is higher if there is a greater amount of this free-type substance. Only a few previous reports have specified the application standards for plasma concentration with a concrete value, but it has been reported that one of the suitable conditions was to have a ≤50% protein-binding rate.

Blood purification falls into roughly 4 categories: hemoperfusion, hemodialysis, hemofiltration and plasma exchange (PE) (table 1). In addition to these, there is a combination of hemodialysis and hemofiltration called continuous hemodiafiltration (CHDF). The removal efficiency of hemoperfusion is not dependent on either the molecular weight or the protein-binding rate of the intoxicating substance. Since the concentration gradient is not utilized for hemoperfusion and the removal of an intoxicating substance can be carried out even at low blood concentrations of the drug, hemoperfusion is usually the first choice for this procedure. Hemoperfusion is especially applicable in those cases where either phenytoin, phenobarbital or theophylline is present. However, its removal efficiency is low for ionized substances and alcohol. Hemodialysis is chosen over hemoperfusion to remove methanol, ethylene glycol, acetylsalicylic acid and lithium [1, 2]. Continuous hemofiltration and CHDF have a smaller cardiovascular impact and are safer to employ on those patients with low blood pressure. With these methods, blood-drug concentration is less likely to rise again after a treatment.

Table 1. Features of blood purification

	Substances removed	Target substances
Hemoperfusion	substances adsorbed on activated carbon	phenytoin; theophylline; phenobarbital
Hemodialysis	MW <2,000; water-soluble substances; V_D <1 l/kg; protein-binding rate: low	alcohol; lithium; acetylsalicylic acid
Hemofiltration	MW <40,000; V_D <1 l/kg; protein-binding rate: low	aminoglycoside
PE	protein-binding rate: high	protein-bound toxins

MW = Molecular weight.

Blood purification is suitable under any of the following conditions:

- systemic conditions becoming progressively severer, even with conservative medical management;
- serious clinical conditions (such as abnormal blood pressure or body temperature, respiratory depression and persistently disturbed consciousness) are observed;
- serious pre-existing disease (such as renal failure, hepatic failure, respiration failure or heart failure);
- organs that usually eliminate toxic substances (such as the kidneys or the liver) are dysfunctional;
- a lethal dose of the intoxicating substance has been ingested.

However, the causative substance may be unclear when the patient is first diagnosed, or, in many cases, several intoxicating substances have been ingested. It is difficult to grasp the volume of distribution and the protein-binding rate of all of the intoxicating substances. Even if the blood-drug concentration of the intoxicating substance can be measured immediately, the obtained measurement is usually the concentration or total amount of plasma, which is not broken down into a classification of bound or free types. In such cases, the results from measurement may not be useful in deciding whether or not blood purification should be implemented. As thus described, fully grasping the disposition of the intoxicating substance and deciding on the suitability of a blood purification therapy is not an easy matter before initiating treatment. Therefore, medical staff should keep in mind that blood purification must be conducted in a timely manner and in accordance with crisis management principle of 'prepare for the worst'.

Table 2. Blood purification for albumin-bound substances

	Sieving coefficient for albumin	Continuous performance	Renal replacement therapy
Plasma exchange	1.0	impossible	impossible
Extracorporeal albumin dialysis			
MARS®	0.01	possible (6–8 h in practice)	good
Prometheus®	0.6	5–6 h	good
PDF	0.3	possible	good

Approach to Protein-Binding Substances

Almost all of the intoxicating substances bind to proteins. There are only a few substances, such as alcohol and lithium, whose protein-binding rates are zero. Paraquat is one example of a substance with a low protein-binding rate. For those intoxication symptoms due to such substances, hemoperfusion or hemodialysis is the best alternative. Should PE be applied in all other intoxication cases? No, the disadvantages of PE should be taken into consideration before it is implemented. PE consumes a large quantity of plasma derivatives and has a high cost. Furthermore, the large membrane pore size eliminates substances essential to the body, including fibrinogen, hepatocyte growth factors, various hormones and insulin. Consequently, the Molecular Adsorbent Recirculating System (MARS®) is used for removal of various intoxicating substances, such as acetaminophen, verapamil and mushrooms [3–5]. This system first isolates the albumin-binding disease agent, and removes the water-soluble disease agent with bicarbonate dialyzing fluid. Then, it absorbs the albumin-binding substance with activated carbon and ion-exchange resin and reuses the albumin suspension as dialyzing fluid (table 2). Recently, Prometheus®, which does not require priming with albumin, has come into use [6]. However, in Japan, albumin dialysis kits such as MARS® and Prometheus® have not been introduced. Thus, we developed plasma diafiltration (PDF) with a guiding concept of selective plasma filtration with dialysis [7–9]. PDF is a means of blood purification that performs PE using a membrane plasma separator with a 0.3 albumin sieving coefficient and streams dialyzing fluid to the outside of the hollow fiber (fig. 1). As mentioned, a variety of methods for removing protein-binding substances while saving plasma derivatives are spreading.

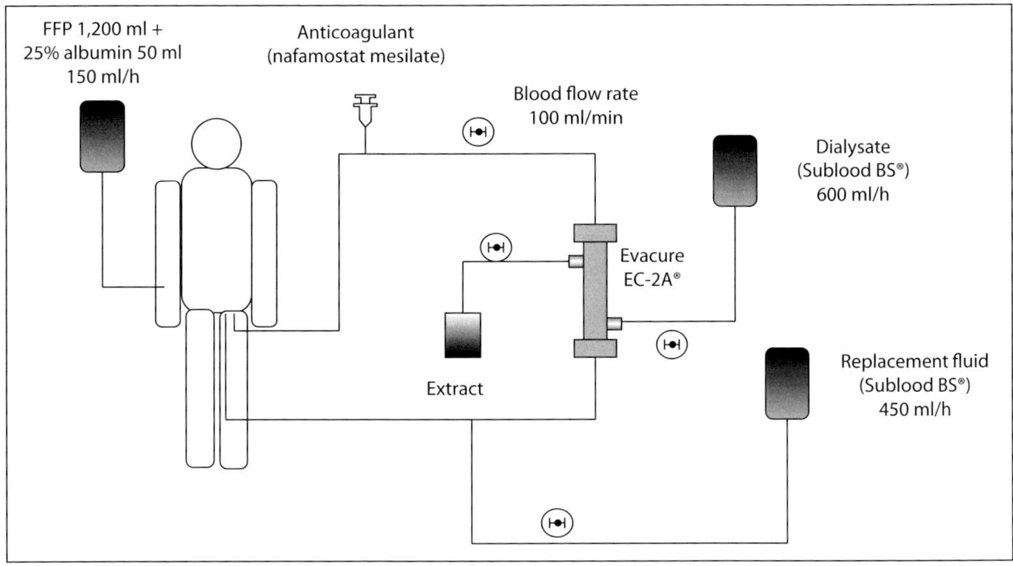

Fig. 1. Schematic representation of PDF. FFP = Fresh frozen plasma. Partly revised from Nakae et al. [7].

Case Presentation

Case of PE Implementation

The patient was a 74-year-old male transported by ambulance. He had ingested mushrooms found in a nearby pine forest, and complained of having palpitations, numbness in his extremities and abnormal vision. He had a history of high blood pressure, and chronic glomerulonephritis after an operation for prostate cancer. Since he was in a state of hysteria and polylogia with high blood pressure (243/96 mm Hg), PE was administered with 2,500 ml of 5% albumin preparation. The process went smoothly and he was able to leave the ICU the next day. The liquid chromatography results revealed that the patient was poisoned with *Amanita ibotengutake*. In 2004, a mass outbreak of *Pleurocybella porrigens* intoxication occurred mainly in Akita, Yamagata and Niigata prefectures. Only those patients with renal failure suffered from intoxication. This case shows those conditions when the patient is under treatment for chronic renal failure and toxic substances easily remain for long periods in the body. For cases of mushroom intoxication in which the bacterial strain cannot be specified, a method of blood purification applicable to a high protein-binding rate should be selected and administered in an aggressive and timely manner.

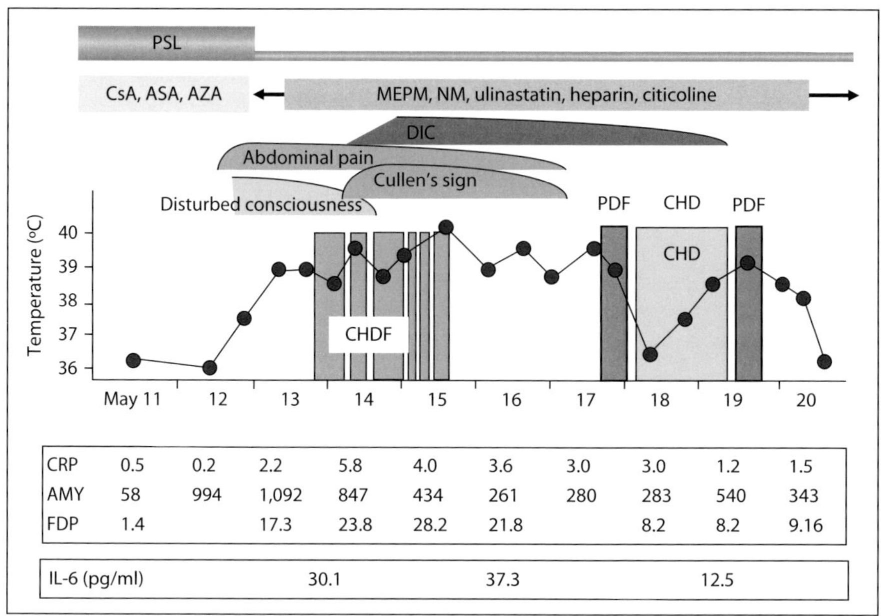

CRP	0.5	0.2	2.2	5.8	4.0	3.6	3.0	3.0	1.2	1.5
AMY	58	994	1,092	847	434	261	280	283	540	343
FDP	1.4		17.3	23.8	28.2	21.8		8.2	8.2	9.16

IL-6 (pg/ml)	30.1	37.3	12.5

Fig. 2. Clinical course (10 days after the onset of acute pancreatitis). PSL = Prednisolone; CsA = cyclosporine A; ASA = 5-aminosalicylic acid; AZA = azathioprine; MEPM = meropenem; NM = nafamostat mesilate; DIC = disseminated intravascular coagulation; CRP = C-reactive protein (mg/dl); AMY = amylase (IU/l); FDP = fibrin degradation product (µg/ml).

Case of PDF Implementation

The patient was a 10-year-old male with superimposed acute pancreatitis and posterior reversible encephalopathy syndrome as side effects of cyclosporine A during treatment for entire colon ulcerative colitis. He had a continuous high fever of 39°C, abdominal pain and rising C-reactive protein levels. Since the results of a CT scan showed that the pancreatitis had become severe, CHDF was started. However, it was interrupted several times by abnormalities in the congealing fibrinogenolysis system. The symptoms did not improve and the development of multiple organ failure became a concern. Therefore, PDF was performed 6 days after onset (fig. 2). His fever went down rapidly after implementation and, after the second PDF, his systemic conditions recovered. The blood IL-6 level decreased after PDF was implemented. PDF is able to selectively remove the albumin-binding substances in the low to middle molecular weight region. On the other hand, coagulation factors can be preserved since the sieving coefficient of fibrinogen is zero [9]. Excess inflammatory cytokines are produced in acute pancreatitis, but these bind with the albumin and are unable to be removed with hemofiltration. A blood purification operation aiming at removing albumin-binding substances such as PDF is an option for cases that develop severe organ failure, as in this case.

Conclusion

The key to blood purification for acute intoxication is the timing of the blood purification implementation and the responses to the protein-binding substances. The timing with which to implement blood purification is comprehensively determined with due consideration for the type of causative substance, the blood concentration and the clinical conditions. An appropriate method to purify blood should be selected according to molecular weight, volume of distribution and protein-binding rate of the causative substance.

References

1 Meertens JH, Jagernath DR, Eleveld DJ, Zijlstra JG, Franssen CF: Haemodialysis followed by continuous veno-venous haemodiafiltration in lithium intoxication: a model and a case. Eur J Intern Med 2009;20:e70–e73.

2 Feinfeld DA, Rosenberg JW, Winchester JF: Three controversial issues in extracorporeal toxin removal. Semin Dial 2006;19:358–362.

3 Koivusalo AM, Yildirim Y, Vakkuri A, Lindgren L, Höckerstedt K, Isoniemi H: Experience with albumin dialysis in five patients with severe overdoses of paracetamol. Acta Anaesthesiol Scand 2003;47:1145–1150.

4 Kantola T, Kantola T, Koivusalo AM, Höckerstedt K, Isoniemi H: Early molecular adsorbents recirculating system treatment of *Amanita* mushroom poisoning. Ther Apher Dial 2009;13:399–403.

5 Nakae H: A case of severe verapamil poisoning treated with molecular adsorbent recirculating system (MARS) (in Japanese). JJOMT 2004;52:321–326.

6 Evenepoel P, Laleman W, Wilmer A, Claes K, Kuypers D, Bammens B, Nevens F, Vanrenterghem Y: Prometheus versus molecular adsorbents recirculating system: comparison of efficiency in two different liver detoxification devices. Artif Organs 2006;30:276–284.

7 Nakae H, Igarashi T, Tajimi K, Kusano T, Shibata S, Kume M, Sato T, Yamamoto Y: A case report of hepatorenal syndrome treated with plasma diafiltration (selective plasma filtration with dialysis). Ther Apher Dial 2007;11:391–395.

8 Nakae H, Igarashi T, Tajimi K, Noguchi A, Takahashi I, Tsuchida S, Asanuma Y: A case of pediatric fulminant hepatitis treated with plasma diafiltration. Ther Apher Dial 2008;12:329–332.

9 Nakae H, Eguchi Y, Saotome T, Yoshioka T, Yoshimura N, Kishi Y, Naka T, Furuya T: A multicenter study of plasma diafiltration (plasma filtration with dialysis) in patients with acute liver failure. Ther Apher Dial, in press.

Hajime Nakae
Department of Emergency and Critical Care Medicine, Akita University Graduate School of Medicine
1–1–1 Hondo
Akita 010-8543 (Japan)
Tel. +81 18 884 6185, Fax +81 18 884 6450, E-Mail nakaeh@doc.med.akita-u.ac.jp

Suzuki H, Hirasawa H (eds): Acute Blood Purification.
Contrib Nephrol. Basel, Karger, 2010, vol 166, pp 100–111

Current Progress in Blood Purification Methods Used in Critical Care Medicine

Akira Saito

Division of Nephrology and Metabolism, Department of Medicine, Tokai University School of Medicine, Isehara, Japan

Abstract

The prognosis of patients with an acute accumulation of pathogenic or toxic substances in their body fluids – a condition that severely affects survival – can be significantly improved by blood purification. The most appropriate blood purification method and the duration for which it should be used must be selected on the basis of efficacy and cost. Several blood purification techniques – such as hemodialysis (HD), hemofiltration (HF), hemodiafiltration, continuous hemofiltration (CHF), hemadsorption and plasma exchange – have been developed. Each modality has different removal capacities and limitations; therefore, it is necessary to thoroughly evaluate the time and the duration of use in the case of different disease conditions. The survival rate of patients treated with HF with 35 ml/min of average filtrate is higher than that observed after conventional HD. In patients with systemic inflammatory response syndrome and multiple organ dysfunction syndrome, proinflammatory cytokines should be removed by HF or CHF, as should the toxins accumulated in the original disease. Thus far, no ideal filter has been developed for the removal of a considerable amount of proinflammatory cytokines with minimal albumin loss. In the case of acute liver failure, ammonia, amino acid metabolites and albumin-binding bilirubin should be removed by a combination of HF and plasma exchange. The use of fresh frozen plasma as a replacement fluid in plasma exchange is also important in order to replenish the deficient coagulation factors and essential metabolic factors. Activation of tissue/organ regeneration by the removal of pathogenic factors or by the substitution of factors essential for regeneration might be important in the case of multiple organ dysfunction syndrome. In critically ill patients with composite conditions, the use of more than two blood purification techniques at the same time or at different times during the course of the diseases can improve patient prognosis more than the use of single methods.

During the Second World War, Kollf et al. [1] succeeded in saving an acute renal failure patient with Crush syndrome using hemodialysis (HD). This was the first time that a renal failure patient was kept alive by treatment using blood purification. However, after that, patients with acute kidney injury (AKI) were only treated with HD until 1960, when Quinton et al. [2] developed an external permanent shunt, by which a patient could be treated with HD repeatedly. Since 1944, blood purification has been used to treat acutely ill patients in whom pathogenic or toxic substances have acutely accumulated endogenously or exogenously, constituting risk factors for survival; blood purification helps in removing these substances and in maintaining homeostasis of body fluids. Therefore, several blood purification methods – such as HD, hemofiltration (HF), hemodiafiltration (HDF), continuous HF and HDF methods (CHF, CHDF), hemadsorption, plasma adsorption, and plasma exchange (PE) – have been developed and used for treating acutely ill patients. With the exception of the adsorption techniques, these have been developed with advances in membrane technology principally based on a dialysis membrane. Blood purification improves the prognosis of acutely ill patients if the progression of the disease cannot be sufficiently slowed by conventional treatments, such as medication, sufficient nutrition and surgery.

In this paper, I have mainly described blood purification methods used in critical care and mentioned the appropriate method for treating several disease conditions.

Application of Blood Purification in Critical Care

Blood purification treatments, in general, are used in emergency rooms and ICU because critically ill patients are treated using total life support, including extracorporeal circulation. The quick removal of toxic/pathogenic substances by blood purification can improve survival rate or at least ensure that the healing time is shortened compared to that required for conventional treatments. Because each method has a distinct technique for the removal of accumulated substances and improves homeostasis in critically ill patients to different extents, the method that is most appropriate for treatment has to be chosen and used in a timely manner, depending on the condition of the patient.

Continuous Blood Purification

High-volume HF has been used for patients with relatively stable hemodynamic status since HF was applied for the treatment of acutely ill patients [3, 4]. For the past 2 decades, on the other hand, CHF and CHDF – which maintain the hemodynamic status at a more stable level than that of intermittent ones – have also been used for the treatment of critically ill patients because of the risks involved with the use of extracorporeal circulation. Compared to intermittent treatment, continuous treatment is beneficial not only because of its effect

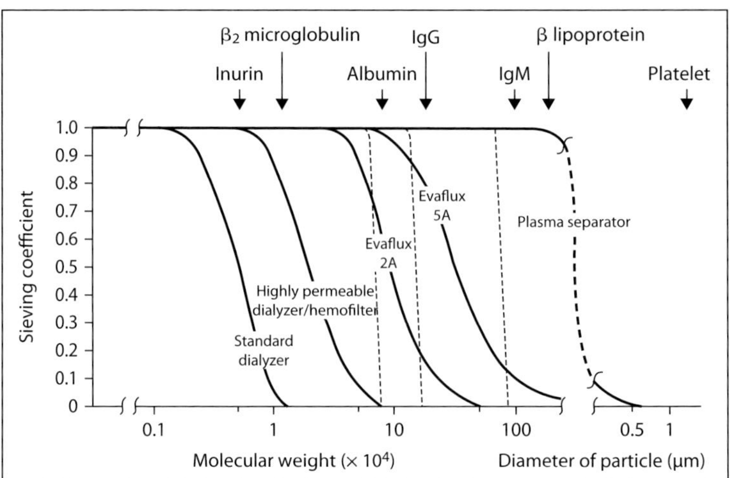

Fig. 1. Five curves for the sieving coefficients of substances with target-molecular-weight ranges filtered across 5 kinds of blood purification membranes: standard dialyzer membrane, highly permeable dialysis/hemofilter membrane, plasma fractionator membranes (such as Evaflux 2A and 5A) and plasma-separator membrane. The sieving coefficients of the target substances were not significantly different from the sieving coefficients of essential substances in the patients. These membranes removed not only toxic and pathogenic substances, but also essential substances to a considerable extent.

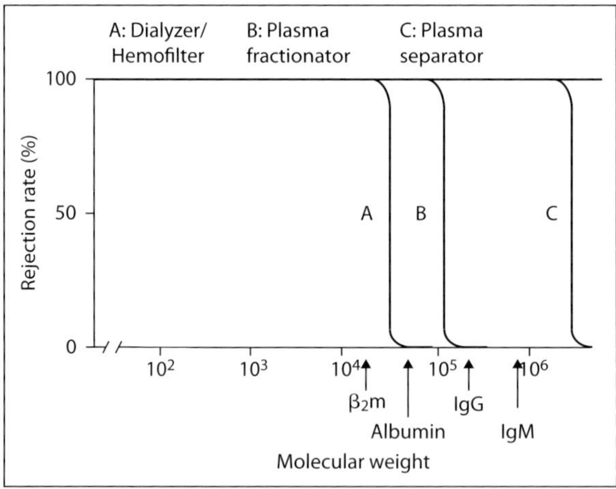

Fig. 2. Ideal blood purification membrane with sharp rejection curves for solutes in 3 molecular-weight ranges. A = β_2-microglobulin to albumin; B = albumin to IgG; C = IgM to platelets (as the smallest blood cells). Improvements should be made in these 3 blood purification membranes to remove target molecules more selectively than conventional membranes. A = Dialysis or HF membrane; B = plasma-fractionation membrane; C = plasma-separation membrane.

on hemodynamic stability but also because of better survival rates and larger total capacity for the removal of the pathogenic and toxic substances and better maintenance of homeostasis in the patients [5, 6]. CHF or CHDF is well used to treat patients with septic shock because of their poor general condition and volume-refractory hemodynamic failure. Patients with severe pancreatitis are also treated mainly with CHF or CHDF because of the hemodynamic instability. CHDF should be selected in the case of patients under severe catabolic stress with remarkably high level of plasma urea nitrogen. HD and HF, on the other hand, are also used in critical-care medicine when there is insufficient manpower to keep patients under continuous observation.

Systemic anticoagulation is essential while using extracorporeal circulation; therefore, in Japan, nafamostat mesilate is used as an anticoagulant instead of heparin in case the patients develop disseminated intravascular coagulation, a bleeding tendency [7], or hypercoagulation due to continuous heparin loading in the continuous blood purification.

Membranes for Blood Purification

It is necessary to use an appropriate blood purification method that can remove the maximum possible amount of accumulated pathogenic substances and maintain maximum homeostasis. Therefore, the characteristics of the membrane – especially the pore size and biocompatibility – are important factors influencing the performance of the HD, HF, CHF and CHDF techniques [8]. The target molecule is different in each disease; therefore, knowledge of its molecular weight, distribution volume and albumin-binding rate is essential when selecting a blood purification device and modality. Sometimes, a patient with composite complications can have more than 2 pathogenic substances. In a blood purification treatment using a highly permeable membrane as a hemodialyzer or a hemofilter, the rejection rate of the target molecule as well as the protein-binding/total ratio of the molecule is important while selecting the most suitable membrane and treatment modality. Sufficiently sharp rejection curves for solutes have never been obtained while using a conventional dialysis membrane, a highly permeable membrane, a plasma-fractionation membrane and a plasma-separation membrane (fig. 1). Through advances in membrane technology, it is now possible to introduce a big difference between the values for the sieving coefficients (SC) of β_2-microglobulin (molecular weight, 11,800 Da) and albumin (molecular weight, 66,800 Da). However, this has not been possible in the case of IL-6 and albumin, though a big difference in SCs between IL-6 and albumin is more important while using CHF and CHDF in critical care medicine. The big difference in the SCs between albumin and immunoglobulin (Ig) G is also important in while using plasma fractionators during PE. Figure 2 shows 3 ideal blood-purification membranes with apparently sharp rejection curves for solutes in 3 different molecular weight ranges. In the figure, A indicates a rejection curve for solutes filtered across an ideal

Table 1. Single and composite blood purification methods used for patients in critical care in the emergency room or ICU of Tokai University School of Medicine from January 1 to December 31, 2005

	Cases (n = 85)
PE	49 (57.6)
HDF	63 (74.1)
Hemadsorption	10 (11.7)
PE, HD	40 (47.1)
PE, HDF	35 (41.2)
PE, CHDF	14 (16.5)
Plasma adsorption, HD, HDF	9 (10.6)
Hemoadsorption, HD, HDF	8 (9.4)

Figures in parentheses are percentages. The patients mentioned in this list received 1 or >2 treatments during different phases of hospitalization.

highly permeable dialysis membrane, B indicates a rejection curve for solutes filtered across an ideal plasma fractionator, and C indicates a rejection curve for solutes filtered across an ideal plasma separator. A shows the big difference in the rejection rates of β_2-microglobulin, IL-6 and albumin; B shows those between albumin and IgG; and C shows those between IgM and platelets as the smallest blood cell.

An alternative to removing target molecules via filtration is the adsorption of cytokine molecules on a membrane. A polymethylmethacrylate membrane filter can adsorb protein molecules, such as β_2-microglobulin, IL-1, IL-6 and tumor necrosis factor, with molecular weights ranging from approximately 10,000 to 40,000 Da; this results in a significant decrease in the plasma levels of these substances, thereby elevating the survival rate of the patients [9].

Combination of Blood Purification Techniques

Combination therapy with 2 or 3 blood purification techniques should be considered in patients with composite disease conditions, such as fulminant hepatitis with AKI, severe pancreatitis with AKI, and septic shock with multiple organ dysfunction syndrome (MODS). Generally, these patients are in a critical condition with severe infectious disease, disseminated intravascular coagulation, gastrointestinal bleeding, symptomatic hypotension, and have an extremely poor nutritional status. Their hemodynamic status has to be

Table 2. Blood-purification methods in critical care medicine, for which the treatment costs have been reimbursed by the Japanese health insurance system

Treatment modality	Applicable disease
HD	AKI
HF	AKI, SAP, ALF
HDF	AKI, SAP, ALF
CHF	AKI, SAP, ALF, endotoxin shock
CHDF	AKI, SAP, ALF, endotoxin shock
Hemoperfusion	ALF, endotoxin shock, SIRS, MODS
PE	ALF, endotoxin shock, SIRS, MODS

SAP = severe acute pancreatitis; ALF = acute liver failure (including fulminant hepatitis and postoperative liver failure); SIRS = systemic inflammatory response syndrome.

carefully maintained during the blood purification treatment. These patients cannot live without total life support, including blood purification. Table 1 shows the blood purification techniques that we used in 2005 at the Kidney Center at Tokai University. In general, the survival rates decreased with an increase in the number of failed organs. Patients with endotoxin shock and AKI were at first treated using an endotoxin-removal column, which improved symptomatic hypotension; this was followed by CHDF for the removal of low-molecular-weight toxins and intermediate-molecular-weight pathogenic substances such as IL-1 and IL-6 and for the normalization of the acid-base and electrolyte imbalances. In contrast, while treating patients with acute liver failure and AKI, high-volume CHF should be applied for the removal of these substances and for maintaining homeostasis; plasma exchange should be used simultaneously for removing protein-bound toxins and high-molecular-weight pathogenic substances.

Characteristics of Blood-Purification Methods Used in Critical Care

At least 8 different blood-purification techniques have been used in critical care medicine. In Japan, treatment costs of some of these techniques have been reimbursed by healthcare insurance in the case of several acute diseases (table 2). The characteristics of each technique are described briefly in the following text.

Hemodialysis

HD, a basic purification modality used to treat patients with AKI or other diseases, removes the accumulated low-molecular-weight substances and/or excess water and normalizes the acid-base imbalance and electrolyte abnormalities. Although conventional HD mainly removes low-molecular-weight substances, HD performed using a highly permeable membrane can remove not only low- but also intermediate-molecular-weight substances such as β_2-microglobulin [10]. HF, which is based on convection, removes intermediate-molecular-weight substances with greater efficacy than HD, which is based on diffusion [11]; however, diffusion removes low-molecular-weight substances to a greater extent than convection in cases where the single filtration volume is less than 80 liters. Recently, HD has again come into the spotlight with regard to blood purification in critical care because slow HD is considered to be safe and effective for use in critical care medicine and because CHF and CHDF are difficult to maintain due to manpower shortage. Highly permeable dialysis membranes have been developed and used to treat patients in critical care, in whom not only low- but also intermediate-molecular-weight toxic or pathogenic substances have accumulated in the plasma.

Hemofiltration

In HF, which is based on convection, the difference between the clearance of low- and intermediate-molecular-weight substances is relatively small when compared with HD; however, the clearance of low-molecular-weight solutes is relatively low while that of the intermediate-molecular-weight solutes is relatively high, compared with HD. In the healthy human kidney, the daily glomerular filtration volume should be approximately 150 liters in order to ensure low plasma concentrations of low-molecular-weight substances such as urea, creatinine, uric acid, ammonia and amino acid metabolites. High-volume HF performed using a highly permeable membrane is recommended for patients whose body fluids contain only intermediate-molecular-weight pathogenic or toxic substances and in patients who have relatively poor hemodynamic stability [12, 13]. Modalities such as CHF and CHDF are recommended in the case of patients with higher concentrations of low- and intermediate-molecular-weight substances and critically poor hemodynamic stability.

Hemodiafiltration

HDF is based on simultaneous diffusion and convection.

During convection, solutes are transported with water flux from the inside to the outside of a porous membrane under hydrostatic pressure from inside or negative pressure from outside the membrane; therefore, dialysate is not required during convection. HDF is a combination of 2 techniques – HD, which removes low-molecular-weight substances to a greater extent, and HF, which removes intermediate-molecular-weight substances to a greater extent.

Thus, HDF is a well-balanced treatment that clears both low- and intermediate-molecular-weight substances to a greater extent than either HD or HF alone.

Continuous Hemofiltration

CHF is useful in the case of critically ill patients with poor hemodynamic stability. Compared to HD, CHF removes intermediate-molecular-weight pathogenic substances to a greater extent and the prognosis also improves significantly [5, 6]. However, when patients develop severe catabolism accompanied by a high plasma urea level, this technique might be insufficient for clearing low-molecular-weight substances. Patients undergoing CHF require enough manpower even if the procedure is performed from daytime to midnight.

Continuous hemodiafiltration

Severely ill patients with systemic inflammatory response syndrome and MODS have frequently been treated with CHDF. The use of this technique leads to a significant decrease in the concentrations of small molecules, such as urea and amino acid metabolites, which accumulate to a remarkable extent in patients with a catabolic condition, and intermediate-molecular-weight substances, including proinflammatory cytokines. In Japan, good results have been obtained when treating patients with poor hemodynamic stability by CHDF at a lower HF rate. The plasma levels of proinflammatory cytokines, such as IL-6, were significantly lowered not with filtration but with adsorption using polymethylmethacrylate membranes [9].

Plasma Exchange

PE has been used to treat patients with acute diseases, such as acute hepatic failure, in which many pathogenic and toxic molecules with widely ranging molecular weights accumulate. Two systems – a centrifugal-exchange method [14] and a membrane-separation method [15, 16] – have been used for plasma exchange. In the past 2 decades, membrane-separation using a plasma separator has been the method of choice for PE in almost all studies.

Single-Filtration Plasma Exchange

Per session, approximately 2.5 l of the patient's plasma is replaced by fresh frozen plasma for a few hours; the pathologic and toxic substances are removed and essential factors such as nutritional and coagulation factors are substituted as fresh plasma. PE is mainly used for treating acute liver dysfunction and diseases that involve accumulation of albumin-bound toxins [17].

Double-Filtration Plasmapheresis

Double-filtration plasmapheresis is mainly used to remove accumulated high-molecular-weight proteins in the patient's plasma – such as antibodies, antigen-antibody complexes, globulins – and other high-molecular-weight substances;

however, it is seldom used for blood purification in critical care. The plasma fraction is separated with the first membrane filter (a plasma separator), and the plasma fraction with a molecular weight less than that of the globulin fraction is separated and returned to blood in patients using the second membrane filter (a plasma fractionator). It results in removal of protein fraction with molecular weight more than albumin. In myasthenia gravis, the patient's conditions are significantly improved by removing considerable amounts of anti-acetylcholine receptor antibody by double-filtration plasmapheresis, or immunoadsorption.

Plasma Diafiltration

In a previous study, I had reported that HDF with a plasma fractionator can be used to treat postoperative hyperbilirubinemia and other diseases wherein there is greater accumulation of intermediate-molecular-weight substances and albumin-bound toxins like bilirubin [18], substituting not only substitution fluid but also considerable amount of fresh frozen plasma. Recently, the same treatment concept was reported as plasma diafiltration in Japan [19]. This blood purification method is considered to be the most effective method for treating patients with pathological conditions where low-molecular-weight substances, albumin-binding substances or higher-molecular-weight substances accumulate. Further clinical trials are required to confirm the efficacy of plasma diafiltration in the treatment of acute liver failure and MODS with hypercytokinemia. The treatment cost has never been reimbursed so far by the Japanese health insurance system.

Adsorption

Direct hemoperfusion and plasma perfusion are used to treat acutely ill patients with drug intoxication, endotoxemia, hyperbilirubinemia, β_2-microglobulin amyloidosis, familial hypercholesterolemia, etc. Direct hemoperfusion with activated charcoal was used to treat patients with acute drug intoxication and hyperbilirubinemia [20–22]. Endotoxin columns with immobilized polymyxin B have been frequently used for patients with endotoxin shock in Japan; this treatment is usually followed by CHDF and results in improved survival rates [23, 24].

Recently, the use of albumin dialysis has been accepted worldwide for the treatment of liver disease and other disease conditions in which there is an accumulation of albumin-bound toxins [25]. The clinical effectiveness of albumin dialysis performed using the molecular adsorbent recirculating system (MARS) and fractionated plasma separation and adsorption system (FPSA/Prometheus) has been reported in the treatment of acute liver diseases such as hepatorenal syndrome [26].

Direct Hemoperfusion

Patients with drug intoxication are treated by HP with activated charcoal as well as with stomach and intestine irrigation and evacuants. Patients with endotoxin shock are treated by HP using polymyxin B immobilized on polystyrene fibers in the perfusion column. Polymyxin B, a neurotoxic and nephrotoxic cationic cyclic decapeptide antibiotic, selectively binds lipopolysaccharide [25, 26]. In Japan, patients with hyperbilirubinemia have been treated using bilirubin adsorption columns with activated charcoal or cationic resin; however, clinical data regarding its significance on prognosis have never been sufficiently confirmed. HP can be used in cases where it is confirmed that coagulation is not stimulated by direct perfusion of blood in the column.

Plasma Perfusion

In plasma perfusion, plasma is first separated using a plasma separator to avoid accelerated coagulation in the column by direct perfusion; following this, the plasma is perfused through the column containing adsorption materials that stimulate blood coagulation. Plasma perfusion with dextran sulfate columns is used for low-density lipoprotein apheresis [27]; immunoadsorption with columns containing immobilized phenylalanine, tryptophan [28] or protein A [29] are used for patients with myasthenia gravis crisis or other autoimmune diseases.

References

1 Kollf WJ, Berk HT, ter Welle M, van der LEY AJ, van Dijk EC, van Noordwijk J: The artificial kidney: a dialyser with a great area. Acta Med Scand 1944;117:121–134.

2 Quinton W, Dillard D, Scribner BH: Cannulation of blood vessels for prolonged hemodialysis. Trans Am Soc Artif Intern Organs 1960;6:104–113.

3 Venkataraman R, Subramanian S, Kellum JA: Clinical review: extracorporeal blood purification in severe sepsis. Crit Care 2003;7:139–145.

4 Schefold JC, Hasper D, Storm C, Corsepius M, Pchowski R, Reinke P: The extracorporeal treatment of septic patients: is there an extrarenal indication? Intensive Med Notfallmed 2007;44:57–63.

5 Ronco C, Bellomo R, Homel P, Brendolan A, Dan M, Piccinni P, La Greca G: Effects of different doses in continuous veno-venous haemofiltration on outcomes of acute renal failure: a prospective randomized trial. Lancet 2000;356:26–30.

6 Saudan P, Niederberger M, De Seigneux S, Romand J, Pugin J, Perneger T, Martin PY: Adding a dialysis dose to continuous hemofiltration increases survival in patients with acute renal failure. Kidney Int 2006;70:1312–1317.

7 Kinugasa E, Akizawa T, Nakashima Y, Wakasa M, Koshikawa S: Nafamostat as anticoagulant for membrane plasmapheresis in high bleeding risk patients. Int J Artif Organs 1992;15:595–600.

8 Haase M, Bellomo R, Morgera S, Baldwin I, Boyce N: High cut-off point membranes in septic acute renal failure: a systematic review. Int J Artif Organs 2007;30:1031–1041.

9 Nakada TA, Oda S, Matsuda K, Sadahiro T, Nakamura M, Abe R, Hirasawa H: Continuous hemodiafiltration with PMMA hemofilter in the treatment of patients with septic shock. Mol Med 2008;14:257–263.

10 Haase M, Bellomo R, Baldwin I, Haase-Fielitz A, Fealy N, Davenport P, Morgera S, Goehl H, Storr M, Boyce N, Neumayer HH: Hemodialysis membrane with a high molecular weight cut-off and cytokine levels in sepsis complicated by acute renal failure: a phase 1 randomized trial. Am J Kidney Dis 2007;50:296–304.

11 Cole L, Bellomo R, Hart G, Journois D, Davenport P, Tipping P, Ronco C: A phase 2 randomized, controlled trial of continuous hemofiltration in sepsis. Crit Care Med 2002;30:100–106.

12 Honoré PM, Zydney AL, Matson JR: High volume and high permeability haemofiltration in sepsis: the evidences and the key issues. Crit Care 2003;3:69–76.

13 Morgera S, Slowinski T, Melzer C, Sobottke V, Vargas-Hein O, Volk T, Zuckermann-Becker H, Wegner B, Müller JM, Baumann G, Kox WJ, Bellomo R, Neumayer HH: Renal replacement therapy with high-cutoff hemofilters: impact of convection and diffusion on cytokine clearances and protein status. Am J Kidney Dis 2004;43:444–453.

14 Yorgin PD, Eklund DK, al-Uzri A, Whitesell L, Theodorou AA: Concurrent centrifugation plasmapheresis and continuous venovenous hemodiafiltration. Pediatr Nephrol 2000;14:18–21.

15 Hirata N, Shizume Y, Shirokaze J, Suemitsu J, Yoshida H, Yamawaki N: Plasma separator Plasmaflo OP. Ther Apher Dial 2003;7:64–68.

16 Sueoka A: Present status of apheresis technologies. 2. Membrane plasma fractionator. Ther Apher 1997;1:135–146.

17 Yamazaki Z, Inoue N: Hepatic assist device, using membrane plasma separator and dialyzer. Med Prog Technol 1987;12:17–24.

18 Saito A, Naito H, Hirohata M: Hemofiltration using filters with high cut-off points in patients with accumulation of molecular substances from 1,000 to 65,000 daltons; in Smeby LC, Jorstad S, Wideroe TE (eds): Immune and Metabolic Aspects of Therapeutic Blood Purification Systems. Basel, Karger, 1986, pp 197–201.

19 Mori T, Eguchi Y, Shimizu T, Endo Y, Yoshioka T, Hanasawa K, Tani T: A case of acute hepatic insufficiency treated with novel plasmapheresis plasma diafiltration for bridge use until liver transplantation. Ther Apher 2002;6:463–466.

20 Kawanshi H, Nishiki M, Sugiyama M, Cho T, Tsuchiya T, Ezaki H: Development of polyetherurethane sheet embedded with powdered charcoal for use in hemoperfusion. Artif Organs 1984; 8:167–173.

21 Nakaji S, Hayashi N: Bilirubin adsorption column Medisorba BL-300. Ther Apher Dial 2003;7:98–103.

22 He NH, Wang YJ, Wang ZW, Liu J, Li JJ, Liu GD, Wang YM: Effects of hemoperfusion adsorption and/or plasma exchange in treatment of severe viral hepatitis: a comparative study. World J Gastroenterol 2004;10:1218–1221.

23 Cruz DN, Perazella MA, Bellomo R, de Cal M, Polanco N, Corradi V, Lentini P, Nalesso F, Ueno T, Ranieri VM, Ronco C: Effectiveness of polymyxin B-immobilized fiber column in sepsis: a systematic review. Crit Care Med 2007;1:R47.

24 Reinhart K, Meier-Hellman A, Beale R, Forst H, Boehm D, Willatts S, Rothe KF, Adolph M, Hoffmann JE, Boehme M, Bredle DL, EASy-Study Group: Open randomized phase 2 trial of an extracorporeal endotoxin adsorber in suspected Gram-negative sepsis. Crit Care Med 2004;32: 1662–1668.

25 Stange J, Mitzner S, Ramlow W, Glieshe T, Hickstein H, Schmidt R: A new procedure for the removal of protein-bound drugs and toxins. ASAIO J 1993;39: M621–M625.

26 Mitzner S, Stange J, Klammt S, Risler T, Erley CM, Bader BD, Berger ED, Lauchart W, Peszynski P, Freytag J, Hickstein H, Loock J, Löhr JM, Liebe S, Emmrich J, Korten G, Schmidt R: Improvement of hepatorenal syndrome with extracorporeal albumin dialysis MARS: results of a prospective randomized, controlled clinical trial. Liver Transpl 2000;6:277–286.

27 Gordon BR, Saal SD. Low density lipoprotein apheresis using the Liposorber dextran sulfate cellulose system for patients with hypercholesterolemia refractory to medical therapy. J Clin Apher 1996;11: 128–131.

28 Konno S, Ichijo T, Murata M, Toda T, Nakazora H, Nomoto N, Sugimoto H, Nemoto H, Kurihara T, Wakata N, Fujioka T: Autoimmune polyglandular syndrome type 2 with myasthenia gravis crisis. Neurologist 2009;15:361–363.

29 Yavuz H, Denizli A: Immunoadsorption of cholesterol on protein A oriented beads. Macromol Biosci 2005;5:39–48.

Akira Saito
Division of Nephrology and Metabolism, Department of Medicine, Tokai University School of Medicine
143 Shimokasuya, Isehara
Kanagawa, 259–1193 (Japan)
Tel. +81 463 93 1121, ext. 2350, Fax +81 463 92 4374, E-Mail asait@is.icc.u-tokai.ac.jp

Suzuki H, Hirasawa H (eds): Acute Blood Purification.
Contrib Nephrol. Basel, Karger, 2010, vol 166, pp 112–118

Membrane Materials for Blood Purification in Critical Care

Akihiro C. Yamashita · Narumi Tomisawa

Department of Materials Science and Engineering, College of Engineering, Shonan Institute of Technology, Kanagawa, Japan

Abstract

Since therapeutic conditions, especially the amount of dialysate, are usually limited in continuous renal replacement therapy (CRRT), selecting an appropriate membrane is more crucial than that in chronic hemodialysis. Under such circumstances, the use of a membrane with adsorption is expected to remove a larger amount of target substances in CRRT. Five commercial dialyzers were investigated to demonstrate the importance of membrane characteristics. The adsorptive characteristics of polymethylmethacrylate (PMMA) membrane were relatively low for cytochrome c (MW 12,400), very strong for α-chymotrypsinogen A (MW 25,000) and relatively strong for albumin (MW 66,000), which may be understood that the adsorption in PMMA has the optimal molecular size. On the other hand, polyacrylonitrile showed relatively low affinity and polysulfone showed essentially no affinity to these protein molecules. Time- and concentration-dependent characteristics of clearance for these proteins were also demonstrated in PMMA. Then we concluded that adsorption found in PMMA may be due to the occlusion of protein molecules into pores of entirely dense membrane. Selecting membrane materials is, therefore, important not only in removing inflammatory cytokines but also in considering the loss of albumin in clinical treatments because even albumin can be adsorbed by the membrane used in blood purification therapies.

Blood purification is one of the most successfully performed artificial organ therapies, with approximately 1.8 million patients being treated worldwide in 2008 [1]. Most of these patients are treated by means of diffusion or ultrafiltration with specifically designed membrane-separation devices. These diffusion or ultrafiltration devices are similar to each other, or sometimes even exactly the same except for the method of usage. When these devices are used in ultrafiltration of blood with substitution (hemofiltration) or dialysis with large amount of ultrafiltration and substitution (hemodiafiltration; HDF), they are called hemofilters or hemodiafilters, respectively. Other than that, these devices are usually called hemodialyzers in which blood and washing solution (dialysate) are separating flowing counter-currently along with the tubular membrane (hemodialysis). These 3 kinds of treatment, i.e., hemodialysis, HDF and hemofiltration were made possible by developing separation membranes especially so-called high-flux membranes that have high hydraulic permeabilities. These membranes are also used for blood purification therapies in critical care [2], sometimes utilizing their adsorption characteristics. Since these treatments are often performed continuously or semi-continuously, they are called continuous renal replacement therapy (CRRT). Continuous HDF (CHDF) is one of the most popular modalities in CRRT. In this study, the importance of membrane characteristics is discussed for 5 commercial dialyzers from the basic mass transport point of view. Adsorption characteristics to protein molecules of these membranes are also discussed for the further success of blood purifications in critical care.

Materials and Methods

In order to specify the importance of membrane characteristics, a series of ultrafiltration experiments were performed with aqueous protein solutions at 310 K for 630 min. An aqueous test solution (2,000 ml; pseudo blood) was prepared with phosphate buffer solution (pH 7.4) in which 1 of the following 3 proteins was dissolved: 0.45 g cytochrome c (MW 12,400, Wako Pure Chemical, Osaka, Japan), 0.90 g α-chymotrypsinogen A (MW 25,000, Sigma-Aldrich, St. Louis, Mo., USA) or 5.10 g bovine serum albumin (MW 66,000, Wako). Five commercial dialyzers (since experiments were ultrafiltration, they are called 'ultrafilters' hereafter) were chosen: CH-1.0N (polymethylmethacrylate, PMMA, arithmetically calculated surface area $A_0 = 1.0\ m^2$); CH-1.0SX (PMMA, $A_0 = 1.0\ m^2$); SH-1.0 (polysulfone, PS, $A_0 = 1.0\ m^2$) – these 3 were produced by Toray Medical, Tokyo, Japan; APS-11S (PS, $A_0 = 1.1\ m^2$); APF-10S (polyacrylonitrile, PAN, $A_0 = 1.0\ m^2$) – these 2 were produced by Asahi Kasei-Kurraray Medical, Tokyo, Japan. Technical specifications of all 5 ultrafilters are listed in table 1.

The blood flow rate (Q_B) was set to 100 ml/min, and the ultrafiltration rate (Q_F) was 10 ml/min. The test solution as well as the ultrafiltrate was returned to the tank, with the expectation of achieving the steady state in which concentrations become unchanged. Samples were taken at various time intervals at the inlet (C_{Bi}) and outlet (C_{Bo}) of the ultrafilter and at the ultrafiltrate (C_F). Absorbance of these samples was directly measured by a spectrophotometer (UV-1600PC, Shimadzu, Kyoto, Japan); wavelengths for cytochrome c, α-chymotrypsinogen A, and albumin were 410, 282 and 278 nm, respectively.

Underlying Theoretical Calculations

The separation performance of solutes by ultrafiltration was characterized by the concept of the sieving coefficient (*s.c.*), which is the ratio of concentration in the downstream to the upstream. However, many definitions may be possible for the *s.c.* when 2

Table 1. Technical specifications of 5 investigated ultrafilters

Brand name	Membrane materials		Surface area, m^2	Effective length, cm	Features	Patients	Maker
	main material	hydrophilic agent					
CH-1.0N	PMMA	none	1.0	19.0	CH-1.0SX has larger pores than CH-1.0N; γ-ray sterilized	acute (CRRT)	Toray Medical
CH-1.0SX			1.0	19.0			
SH-1.0	PS	polyvinyl-pyrrolidone	1.0	19.0	γ-ray sterilized		
APS-11S			1.1	23.7		chronic dialysis	Asahi Kasei-Kurraray
APF-10S	PAN	acrylic acid	1.0	15.2	ETO sterilized	acute (CRRT)	

parallel currents are present along the membrane. We used the following simple definition [3, 4], $s.c._4$, which produces a close value to one from another rigorous definition in which the concentration distribution and flow effects in a module are taken into account:

$$s.c._4 = \frac{C_F}{\sqrt{C_{Bi} \times C_{Bo}}} \tag{1}$$

where C_{Bi} and C_{Bo} are solute concentrations of the test solution at the inlet and outlet of the ultrafilter, respectively (mg/ml), and C_F is the solute concentration of the ultrafiltrate (mg/ml).

In ultrafiltration experiments under total recirculation of the test solution and ultrafiltrate, one should expect the steady state over time in which C_{Bi}, C_{Bo}, and C_F are constant; however, before the steady state is achieved, trapping (adsorption) by the membrane may occur. Under such circumstances, clearances C_L (ml/min) are used as an index to evaluate how much solute could be removed by adsorption in addition to ultrafiltration:

$$C_L = \frac{C_{Bi} - C_{Bo}}{C_{Bi}} Q_{Bo} + Q_F \tag{2}$$

where Q_{Bo} (ml/min) is the blood flow rate at the outlet of the dialyzer ($= Q_{Bi} - Q_F$). Also the decreased concentration in the test solution *(C_{Bi})* should correspond to the amount of adsorption. Then the following equations calculate the amount of adsorbed proteins M_{ads} (mg) and the fractional adsorption M_{ads}/M_∞ (no unit) during the course of experiment, respectively:

$$M_{ads} = V \cdot \{C_{Bi}(0) - C_{Bi}(t)\} \tag{3}$$

$$\frac{M_{ads}}{M_\infty} = \frac{V \cdot \{C_{Bi}(0) - C_{Bi}(t)\}}{V \cdot C_{Bi}(0)} = \frac{C_{Bi}(0) - C_{Bi}(t)}{C_{Bi}(0)} \tag{4}$$

where $C_{Bi}(0)$ and $C_{Bi}(t)$ are concentrations of solute of interest in the test solution at initial ($t = 0$) and at arbitrary time 't', respectively, V is the volume of the test solution (ml) and M_∞ is the initial amount of the solute in the test solution (mg).

Results and Discussion

In terms of $s.c._4$ for cytochrome c, although two PMMA ultrafilters showed a little lower steady-state values (approximately 0.90) than other 3 ultrafilters (approximately 0.95), no other big differences were found among them (data not shown). Therefore, we would expect little difference in clinical treatments, no matter which of these 5 ultrafilters were used when removing solutes similar to cytochrome C in size. Figure 1 shows $s.c._4$ for α-chymotrypsinogen A in 5 ultrafilters. Values of $s.c._4$ at steady state were around 0.9, which was almost the same as the ones found for cytochrome C, although the molecular weight is twice as large in α-chymotrypsinogen A. More importantly, however, two PMMA ultrafilters showed different behaviors compared with other 3 ultrafilters, showing much slower increases in $s.c._4$ values from 0 to 0.9. It also took 480 min to achieve a plateau from the beginning of the experiment. This phenomenon is explained by much slower appearance of the protein molecule in the ultrafiltrate than its elimination from the blood compartment, which is usually called adsorption. This was verified by plotting the clearances calculated from the same data in figure 2. The initial high clearances in PMMA are caused by adsorption while the stable values (~9 ml/min) were caused by ultrafiltration that is a product of $s.c._4$ and Q_F or $0.9 \times 10 = 9.0$ ml/min.

As previously mentioned, since $s.c._4$ for cytochrome c in 5 ultrafilters were similar to each other, no significant adsorption may be expected for substances similar (physically and/or chemically) to cytochrome c. However, since totally different curves were obtained in $s.c._4$ and in clearance between PMMA and other ultrafilters in ultrafiltration of α-chymotrypsinogen A, adsorption in PMMA may play a significant role in removing molecules that are similar to α-chymotrypsinogen A in size. According to a classic textbook [5], there are 3 major mechanisms for physical adsorption (physisorption): the van der Waals force, electrostatic effect and hydrophobic interaction. The van der Waals force,

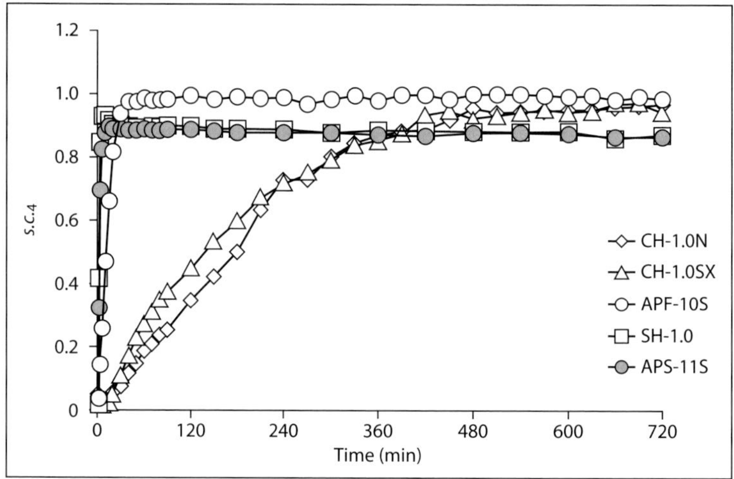

Fig. 1. Time courses of the $s.c._4$ for α-chymotripsinogen A (MW 25,000) in five ultrafilters. Two ultrafilters with PMMA membrane. One ultrafilter with PAN membrane. Two ultrafilters with PS membrane.

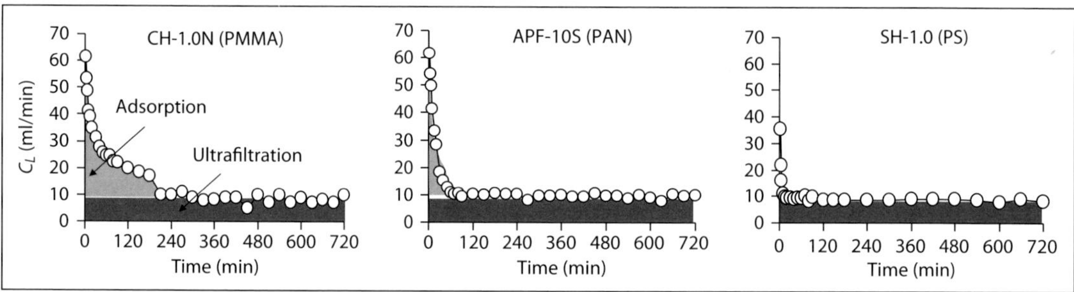

Fig. 2. Time courses of clearance for α-chymotripsinogen A in three ultrafilters. The initial high clearances in PMMA are caused by adsorption while the stable value (~9 ml/min) is caused by ultrafiltration that is the product of $s.c._4$ and Q_F or 0.9 × 10 = 9 ml/min. Not much and essentially zero adsorption were found in PAN and in PS, respectively.

however, is usually known to have relatively small effect and the other 2 mechanisms should not be dependent on the molecular weight. So, the adsorption found between PMMA and α-chymotrypsinogen A must be caused by a somewhat different mechanism. Since the PMMA membrane has an entirely dense structure [6], the sorption may be the occlusion of protein molecules in specific size into pores of the membrane. If so, the adsorption is dependent on the molecular size and there exists certain range of molecular sizes that effectively cause the adsorption.

Table 2. Amount and fractional adsorption of albumin by 5 commercial ultrafilters

	CH-1.0N	CH-1.0SX	SH-1.0	APS-11S	APF-10S
Main material of the membrane	PMMA		PS		PAN
Fractional adsorption rate (over 10.5 h), %	20	50	4.4	5.5	5.0
Adsorbed albumin (over 10.5 h), g/m^2	1.02	2.55	0.22	0.25	0.26

Clearances for inflammatory cytokines, especially for IL-6, have been reported to show correlations with their concentrations in CHDF with PMMA [7]. As shown in figure 2, higher clearances are expected at the beginning of the treatment while the concentration is still high. Therefore, figure 2 may explain the correlation between clearance and concentration, although the concentrations of these cytokines do not drastically change over time in clinical situations.

The amount of adsorption and fractional adsorption of albumin by 5 commercial ultrafilters are tabulated in table 2. The amount of albumin by PS and PAN membranes was 0.22–0.26 g/m^2, while those for PMMA was 1.02 and 2.55 g/m^2, respectively, for CH-1.0N and CH-1.0SX. Therefore, PMMA has 4–12 times higher adsorption characteristics for albumin than other membranes. Moreover, the larger the pore size, the higher the amount of albumin adsorption was found, which also suggests an occlusive mechanism of adsorption of albumin into the pores of the membrane. Since PMMA has still high adsorption characteristics to albumin, one should pay a great deal of attention to the albumin loss not only by permeation across the membrane but also by adsorption by the membrane. These facts also suggest that the adsorption in PMMA has the optimal molecular size into dense structures of the membrane.

Many reports in mid 1980s suggested that selecting a biocompatible membrane offered many advantages to patients with acute renal failure; however, whether biocompatibility issues would influence outcome was unclear and remained to be proven [8]. The amount of dialysate has been extensively studied since then [9, 10], including whether or not dialysis is necessary in addition to ultrafiltration [11–13]. However, little has been noted on the membrane material for acute renal failure. According to the data demonstrated in figures 1 and 2, selecting membrane materials is even more crucial in blood purifications in critical care, mainly because of the adsorption characteristics of the membrane materials. Under limited therapeutic conditions, since it is difficult to achieve much solute removal by diffusion or ultrafiltration, adsorption should play a significant role in removing toxic substances.

Conclusions

PMMA membrane showed much higher adsorption characteristics to protein molecules than PS and PAN membranes. Adsorption in PMMA has optimal molecular size and may be due to occlusion into the pores, which should be taken into account when determining albumin loss in clinical treatment. Selecting membrane materials may be more important in blood purification in critical care than that in chronic dialysis because adsorption plays significant roles in the former.

References

1 Fresenius Medical Care: ESRD Patients in 2008 – A Global Perspective. Bad Homburg, Fresenius Deutschland, 2009, p 4.
2 Nissenson AR: Acute renal failure: definition and pathogenesis. Kidney Int 1998;53(suppl 66):S7–S10.
3 Yamashita AC, Sakiyama R, Hamada H, Tojo K: Two new definitive equations of the sieving coefficient (in Japanese). Kidney Dial (Jin To Toseki) 1998;45:S36–S38.
4 Yamashita A: New dialysis membrane for removal of middle molecule uremic toxins, Amer J Kid Dis 2001;38(suppl 1):S217–S219.
5 Glsstone S, Lewis D: Elements of Physical Chemistry, ed 2. Princeton, D van Nostrand, 1960, pp 558–572.
6 Sakai Y, Tsukamoto H, Fujii Y, Tanzawa H: Formation of poly(methyl methacrylate) membranes utilizing stereocomplex phenomenon; Cooper AR (ed): Ultrafiltration Membranes and Applications. New York, Plenum, 1980, pp 99–107.
7 Matsuda K, Hirasawa H, Oda S, Shiga H, Nakanishi K: Current topics on cytokine removal technologies. Ther Apher 2001;5:306–314.
8 Aoike I: Clinical significance of protein adsorbable membranes – long-term clinical effects and analysis using a proteomic technique. Nephrol Dial Transplant 2007;22(suppl 5):v13–v9.
9 Birk HW, Kistner A, Wizemann V, Schutterle G: Protein adsorption by artificial membrane materials under filtration conditions. Artif Organs 1995;19:411–415.
10 Kurtal H, Herrath DV, Schaefer K: Is the choice of membrane important for patients with acute renal failure requiring hemodialysis? Artif Organs 1995;19:391–394.
11 Ronco C, Ricci Z, Homel P, Brendolan A, Dan M, Piccinni P, Greca GL: Effects of different doses in continuous veno-venous hemofiltration on outcomes of acute renal failure: a prospective randomized trial. Lancet 2000;355:26–30.
12 Joannes-Boyau O, Rapaport S, Bazin R, Fleureau C, Janvier G: Impact of high volume hemofiltration on hemodynamic disturbance and outcome during septic shock. ASAIO J 2004;50:102–109.
13 Saudan P, Niederberger M, Seignwux SD, Romand J, Pugin J, Perneger T, Martin PY: Adding a dialysis dose to continuous Hemofiltration increases survival in patients with acute renal failure. Kidney Int 2006;70:1312–1317.

Prof. Akihiro C. Yamashita, PhD
Department of Materials Science and Engineering, College of Engineering, Shonan Institute of Technology
1-1-25 Tsujido-Nishikaigan, Fujisawa
Kanagawa 251-8511 (Japan)
Tel./Fax +81 466 30 0234, E-Mail yama@la.shonan-it.ac.jp

Suzuki H, Hirasawa H (eds): Acute Blood Purification.
Contrib Nephrol. Basel, Karger, 2010, vol 166, pp 119–125

Anticoagulation in Acute Blood Purification for Acute Renal Failure in Critical Care

Toshio Shinoda

Dialysis Center, Kawakita General Hospital, Tokyo, Japan

Abstract

The correct selection of anticoagulation in acute blood purification is crucial for avoiding exacerbation of bleeding in critical care patients with acute renal failure, as these patients frequently exhibit hemorrhagic complications. The mode of acute blood purification is determined mainly by the patient's hemodynamic stability, and continuous renal replacement therapies (CRRTs) have been extensively performed for patients with hemodynamic instability. Unfractionated heparin, low molecular weight heparin and nafamostat mesilate (nafamostat) are available in acute blood purification for the patients. Special caution should be taken when using either type of heparin in CRRT because of their antithrombin effect, long half life and large dose, and the prolonged treatment time of CRRT. This is especially the case with patients of small stature, which is the case for many Japanese people. Nafamostat can be used safely in CRRT for critical care patients with acute renal failure and bleeding risks, because it acts as a regional anticoagulant due to its pharmacological characteristics. Nafamostat has been widely used in acute blood purification at critical care units in Japan.

Many critical care patients with organ failures, intoxication or refractory medical diseases are extensively treated with acute blood purification therapies today. Acute renal failure (ARF) is most frequent among these diseases, and is often complicated by multiple organ dysfunction and/or bleeding lesions. Anticoagulation is essential for the extracorporeal circulation in blood purification therapies. Thus, for patients with hemorrhagic complications and/or bleeding tendency and those who are post-operative it is crucial to select the correct anticoagulation if exacerbation of bleeding is to be avoided.

This is especially the case with the continuous therapy mode, which is mainly indicated for patients with circulatory failure, because the treatment time is

Table 1. Indication for acute blood purification in critical care patients

With bleeding risk	Without bleeding risk
ARF after surgery after trauma due to hemorrhagic shock due to DIC MODS with bleeding tendency complicated by bleeding lesions Hepatic failure post surgery Severe hepatic failure with bleeding tendency Severe acute pancreatitis	Other ARF MODS without bleeding tendency Fulminant hepatitis without bleeding tendency Systemic inflammatory response syndrome Intoxication Refractory medical diseases (e.g. congestive heart failure, TTP/HUS, immune mediated neurological diseases, toxic epidermal necrolysis)

DIC = disseminated intravascular coagulation; HUS = hemolytic uremic syndrome; MODS = multiple organ dysfunction syndrome; TTP = thrombotic thrombocytopenic purpura.

prolonged for more than 12 h. I describe here the anticoagulation methods in acute blood purification, focusing on patients at risk for bleeding.

Indications for Acute Blood Purification in Critical Care Patients

Acute blood purification therapies are mainly applied in ARF, multiple organ dysfunction syndrome, systemic inflammatory response syndrome, hepatic failure, and refractory medical diseases such as congestive heart failure and immune-mediated disorders (table 1). Critical care patients with ARF frequently exhibit hemorrhagic complications. Such conditions include ARF after surgery, ARF after trauma, ARF due to hemorrhagic shock, ARF due to disseminated intravascular coagulation and ARF representing multiple organ dysfunction syndrome with bleeding tendency. Proper anticoagulation should be selected in order to avoid exacerbation of bleeding during the acute blood purification therapy for these diseases. Anticoagulation methods in other diseases without bleeding risk have been established, and special caution is not necessarily needed.

Modalities in Acute Blood Purification for ARF and Their Anticoagulation

Today there are 3 major modalities in acute blood purification for ARF: the continuous mode, the intermittent mode, and the extended daily mode (table 2).

Table 2. Acute blood purification for ARF

Mode	Blood flow rate, ml/min	Treatment time, h
CRRT	50–120	12–24 using small devices
IRRT	200–400	3–5
Extended daily hemodialysis	200 (100–150 in patients with circulatory instability)	6–8

The continuous mode is mainly indicated for patients with ARF whose hemodynamic state is disturbed. It has been called continuous renal replacement therapy (CRRT). CRRT consists of continuous hemodialysis (CHD), continuous hemodiafiltration (CHDF) and continuous hemofiltration (CHF), and has been extensively performed in critical care [1–3]. These therapies apply also to critical care patients without ARF, in which case CRRT is considered for nonrenal indications. CHF is a modified mode of continuous arterivenous hemofiltration [4], the prototype of CRRT, using a blood pump in the extracorporeal circuit. CHD and CHDF are other modified modes of continuous arterivenous hemofiltration. CHF, CHD and CHDF are usually performed continuously at a low blood flow rate (50–120 ml/min) for 24 h (or 12 h at minimum) [5].

In contrast, the standard therapies of hemodilysis, hemodiafiltration and hemofiltration are now termed intermittent renal replacement therapy (IRRT), in the same manner as CRRT. These therapies are usually performed at a high blood flow rate (200–400 ml/min) for 3–5 h.

Both CRRT and IRRT are applied to critical care patients with ARF. A randomized controlled trial [6] did not demonstrate a difference in survival between ARF patients treated with IRRT and those treated with CRRT. It demonstrated, however, that CRRT had more favorable effect on hemodynamic state than IRRT. It is worth noting that the study used blood flow rates of 300 ml/min in IRRT and 200 ml/min in CRRT. These high blood flow rates might worsen the survival in patients with circulatory instability in both treatment groups. Another possibility is that the survival rate might depend rather on the underlying disease than on the treatment mode. These are presumably the reasons why there was no difference in survival between the 2 treatment modes. Thus CRRT has been extensively applied to critical care patients with ARF whose hemodynamic state is disturbed.

The extended daily mode has been proposed as an intermediate treatment mode between CRRT and IRRT. Extended daily hemodialysis is performed at an intermediate blood flow rate (200 ml/min) for 6–8 h, usually during daylight hours [7]. The blood flow rate could be less (100–150 ml/min) for patients with hemodynamic instability.

In case of using heparin, patients treated with CRRT might be more prone to bleeding than those treated with IRRT or extended hemodialysis, because the

Table 3. Anticoagulation in acute blood purification for ARF

Drugs	Dose for patients with bleeding risks	Dose monitoring
UF heparin	300–500 IU/h (500 IU for the initial dose or priming)	ACT, APTT
LMW heparin	200–350 IU/h (250 IU for the initial dose or priming)	(anti-Xa activity)[a]
Nafamostat mesilate	20–50 mg/h (20–50 mg for priming)	ACT

ACT = activated coagulation time; APTT = activated partial thromboplastin time.
[a] Not measurable at bedside.

total amount of heparin becomes larger in CRRT due to its prolonged treatment time than in the other modes. It is favorable to change the treatment mode from CRRT to extended daily hemodialysis or IRRT when the patient's hemodynamic state improves in order to reduce the bleeding risk.

Anticoagulation in Acute Blood Purification for ARF

In acute blood purification for critical care patients with ARF, 3 kinds of anticoagulation have been available (table 3). In addition to standard unfractionated (UF) heparin and low molecular weight (LMW) heparin, nafamostat mesilate (nafamostat) is a unique anticoagulation agent which was developed for use in hemodialysis patients with bleeding risks [8, 9], although it has not been available in Western countries. Special caution is needed not to exacerbate bleeding when using UF or LMW heparin in acute blood purification for critical care patients with bleeding risks.

Unfractionated Heparin

UF heparin inhibits activated coagulation factors II, IX, X, XI, XI and XII (thrombin, IXa, Xa, XIa and XIIa), and its anticoagulation action depends mainly on antithrombin and anti-Xa effects (fig. 1). Fibrin formation is inhibited by its antithrombin effect, and hemostasis is accordingly disturbed by lack of fibrin thrombus formation.

It was reported that a low dose heparin (300–500 IU/h) was available for anticoagulation in CRRT for critical care patients with ARF and bleeding risks

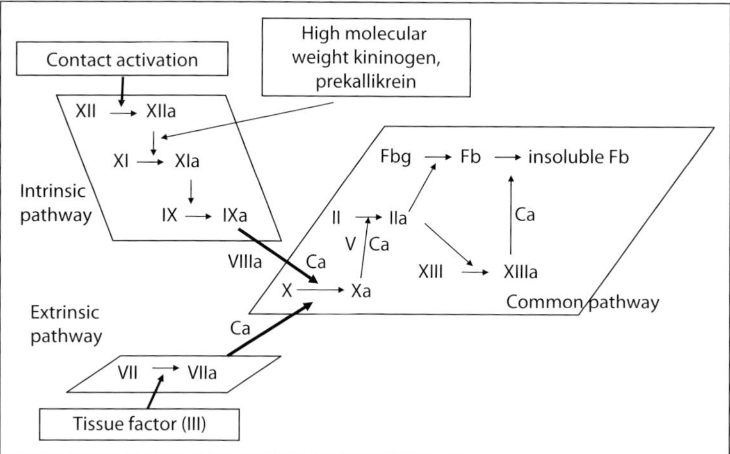

Fig. 1. Coagulation pathway.

[5]. The dose could be adjusted according to the result of activated coagulation time (ACT) or activated partial thromboplastin time (APTT) at the bedside. The optimal value of these tests in the sample at the venous circuit is 1.5–2 times the baseline value before starting the acute blood purification therapy. Nevertheless, special caution should be taken to avoid bleeding because the total dose becomes large in CRRT as the therapy is performed for 12–24 h. The estimated amount reaches 3,600–12,000 IU per session. Bleeding might be exacerbated by the inhibition of hemostasis due to its potent antithrombin effect.

CRRT is performed at a low blood flow rate using a small size hemofilter/hemodiafilter and blood circuit. The methodological characteristics have 2 opposite influences on the heparin dosage.

One is considered to be a possible adverse influence. The low blood-flow rate of CRRT requires higher heparin dosage to prevent coagulation in the extracorporeal circuit. This is because the extracorporeal blood is in contact with the surface of the blood circuit and the hemofilter/hemodiafilter for a longer period, according to the lower shear rate in the extracorporeal circuit.

Another is a favorable influence. The heparin concentration in the extracorporeal circulation of CRRT becomes higher than in that of IRRT even when the same dose is applied to the extracorporeal circuit, because of its smaller priming volume [5]. This makes the heparin dose lower to achieve anticoagulation in the circuit, and the systemic heparin concentration resultantly becomes lower in CRRT than that in IRRT. This is favorable to avoid exacerbation of bleeding in CRRT with heparin. The latter influence is actually considered to overcome the above-mentioned adverse one.

However, the favorable influence becomes less in patients of small stature, as is the case for many Japanese people, because the required dose of heparin for the extracorporeal circulation of CRRT is mainly determined not by the body size but by the priming volume of the extracorporeal circuit. When the body weight is two thirds (for example 50 vs. 75 kg), the heparin concentration in the systemic circulation would be 1.5 times. UF heparin usage in CRRT for critical care patients of small stature might possibly exacerbate bleeding.

LMW Heparin

LMW heparin is also available in acute blood purification for critical care patients with ARF and bleeding risks. LMW heparin is lower molecular fractions of UF heparin, and possesses a small antithrombin effect. Another characteristic of LMW heparin is its longer half life (2–3 h) than that of UF heparin (1–1.5 h).

LMW heparin has been reported to be a safe anticoagulant for hemodialysis patients with bleeding risk [10]. However, a randomized controlled trial [11] did not demonstrate the superiority of LMW heparin to UF heparin. This lack of difference is considered to arise from the fact that there was no bleeding complication in either the UF or LMW heparin groups.

Because of the longer half life of LMW heparin, the dose of LMW heparin required in hemodialysis/hemofiltration is reported to be a half for the initial dose and two thirds for the continuous infusion [11]. Thus, LMW heparin at a low dose (200–350 IU/h) could also be available for anticoagulation in CRRT. Exacerbation of bleeding is considered to be less, because the antithrombin effect of LMW heparin is much weaker and the dose is less than UF heparin. Special caution should also be taken when using LMW heparin in CRRT because of its longer half life than UF heparin. Another caution for using LMW heparin in CRRT is that the dose monitoring cannot be conducted by bedside coagulation tests such as ACT and APTT, because of its very weak antithrombin effect.

Nafamostat Mesilate

Nafamostat is a synthetic serine protease inhibitor which inhibits activated coagulation factors thrombin, Xa and XIIa, kallikrein, and plasmin [12] in addition to platelet [8]. The early components of the intrinsic coagulation pathway, such as XIIa and kallikrein, are more potently inhibited than the late components [12].

It is considered that nafamostat acts as a regional anticoagulant because of its very short anticoagulation effect. In addition, about 40% of nafamostat is removed by dialysis and/or convection in the extracorporeal circuit and is then rapidly degradated by esterases in the liver and blood [9]. Thus, APTT values at the arterial circuit (before nafamostat infusion) were prolonged a little while those at the venous circuit were prolonged more than 2-fold of the baseline values in hemodialysis patients using nafamostat. The result indicates that the anticoagulation effect of nafamostat is almost completely inactivated in the systemic circulation. ACT is reportedly not prolonged at 15 min after the completion of hemodialysis in case of using nafamostat as anticoagulant [8].

It is accordingly considered that nafamostat can be use safely in CRRT for critical care patients with ARF and bleeding risks at the dose of 20–50 mg/h [8]. The dose could be adjusted according to the result of ACT at the bedside. The optimal value of the test in the sample at the venous circuit 1.5–2 times the baseline value. Nafamostat has been widely used in acute blood purification in critical care units in Japan [13].

References

1 Caplan AA: Continuous renal replacement therapy (CRRT) in the intensive care unit. J Intensive Care Med 1998;13:85–105.

2 Ronco C, Bellomo R, Ricci Z: Continuous renal replacement therapy in critically ill patients. Nephrol Dial Transplant 2001;16(suppl):S67–S72.

3 Oda S, Hirasawa H, Shiga H, Nakanishi K, Matsuda K, Nakamura M: Continuous hemofiltration/hemodiafiltration in critical care. Ther Apher 2002;6:193–198.

4 Kramer P, Schrader J, Bohnsack W, Grieben G, Grone HJ, Scheler F: Continuous arteriovenous hemofiltration: a new kidney replacement therapy. Proc Eur Dial Transplant Assoc 1981;18:743–749.

5 Ronco C: Continuous renal replacement therapies for the treatment of acute renal failure in intensive care patients. Clin Nephrol 1993;40:187–198.

6 Augustine JJ, Sandy D, Seifert TH, Paganini EP: A randomized controlled trial comparing intermittent with continuous dialysis in patients with ARF. Am J Kidney Dis 2004;44:1000–1007.

7 Kumar VA, Craig M, Depner TA, Yeun JY: Extended daily dialysis: a new approach to renal replacement for acute renal failure in the intensive care init. Am J Kidney Dis 2000;36:294–300.

8 Akizawa T, Koshikawa S, Ota K, Kazama M, Mimura N, Hirasawa Y: Nafamostat mesilate: a regional anticoagulant for hemodialysis in patients at high risk for bleeding. Nephron 1993;64:376–381.

9 Matsuo T, Kario K, Nakao K, Yamada T, Matsuo M: Anticoagulation with nafamostat mesilate, a synthetic protease inhibitor, in hemodialysis patients with a bleeding risk. Haemostasis 1993;23:135–141.

10 Koshikawa S, Akizawa T, Mimura N, Ota K, Hirasawa Y, Maekawa M: Effects of low molecular weight heparin as an anticoagulant for hemodialysis patients with bleeding risk: multi-institutional control study with low dose or regional heparinization. Clin Eval 1991;19:541–571.

11 Schrader J, Stibbe W, Armstrong VW, Kandt M, Muche R, Kostering H, Seidel D, Cheler F: Comparison of low molecular weight heparin to standard heparin in hemodialysis/hemofiltration. Kidney Int 1988;33:890–896.

12 Hitomi T, Ikari N, Fujii S: Inhibitory effect of a new synthetic protease inhibitor (FUT-175) on the coagulation system. Haemostasis 1985;15:164–168.

13 Ohtake Y, Hirasawa H, Sugai T, Oda S, Shiga H, Matsuda K, Kitamura N: Nafamostat mesilate as anticoagulant in continuous hemofiltration and continuous hemodiafiltration. Contrib Nephrol 1991;93:215–217.

Toshio Shinoda, MD, PhD
Dialysis Center, Kawakita General Hospital
1-7-3 Asagaya-kita, Suginami-ku
Tokyo 166-8588 (Japan)
Tel. +81 3 3339 2121, Fax +81 3 3339 2986, E-Mail tshinoda@kawakita.or.jp

Suzuki H, Hirasawa H (eds): Acute Blood Purification.
Contrib Nephrol. Basel, Karger, 2010, vol 166, pp 126–133

Equipment and Monitoring in Continuous Renal Replacement Therapy

Yoshihisa Yamashita[a] · Isao Tsukamoto[b] · Yoshihiko Kanno[c] · Hiromichi Suzuki[b,c]

[a]School of Biomedical Engineering, Faculty of Health and Medical Care, [b]Department of Medical Engineering, Saitama International Medical Center, [c]Department of Nephrology, School of Medicine, Faculty of Medicine, Saitama Medical University, Saitama, Japan

Abstract

Continuous renal replacement therapy is expected to improve unfavorable status in critical care. As precise volume control is most important to maintain the damaged circulation system, blood and solution control should be carried out precisely. Recently, further technical development was achieved in this area, and quality of these products – including disposable kits – has been improved. Nevertheless, incidental and accidental errors in human and equipment are sometimes happen. In order to decrease it, the staff understands the system of treatment and is familiar to the routine check point.

Continuous renal replacement therapy (CRRT) was established in 1977 by Kramer et al. [1], who controlled the body fluid balance using hemofiltration technology by the difference in arterial and venous pressures.

Because severely ill patients should be monitored throughout therapy, it is necessary to share information among the professional staff to foster good cooperation. Moreover, durability and safety in continuous usage are very important considerations when selecting the console system rather than the usual intermittent hemodialysis treatment. We describe the console and monitoring system used in CRRT to provide safe management.

Selection of the Console System in CRRT

To control the unstable cardiorespiratory system in critical patients, many monitoring and therapeutic devices are often set up in close proximity. Medical

Table 1. Safety features of the CRRT console

Simple operation
Distinctly visualized
Compact and easily movable
Precisely controlled pump system
Complete alarm system
Endurance and long life

engineers frequently find it difficult to secure enough space to set up the CRRT console system. In such a situation, it is better for medical staff to carry out CRRT around the bed without a complex tubal system, unlike the standard intermittent hemodialysis unit which needs heavy devices and requires skillful handling. For safety of therapy, CRRT console devices must meet the conditions outlined in table 1.

Recently, CRRT console devices have improved so that, in addition to dialysis and filtration, they have various functions for other hemopurifications, such as apheresis by changing the extra-corporeal blood system. For the selection of the CRRT console, it is important to consider many factors, such as safety, flexibility and operation.

Composition and Monitoring of the CRRT Console

A CRRT console system is usually composed of pump systems, pressure monitors, and a flow controller, each for blood, fresh and spent dialysate, substitution, filtrate and anti-coagulation (fig. 1 and table 2). It is also expected to satisfy international standard EMC (IEC-60601-1-2) concerning electromagnetic fields. As an extra-corporeal blood tube is often manufactured adjusted to each console system, staff can set and start the priming procedure quickly and easily. The priming volume is usually 50–90 ml for the adult system.

Blood Flow Rate (Qb)

A double lumen catheter is usually selected as vascular access in continuous blood purification with pump systems composed by 2–4 roller type pumps to maintain stable blood flow in the circulation system. The main reason for difficulty with blood flow is inflow failure caused by the patient moving their body. When blood flow is reduced for any reason, the console system will

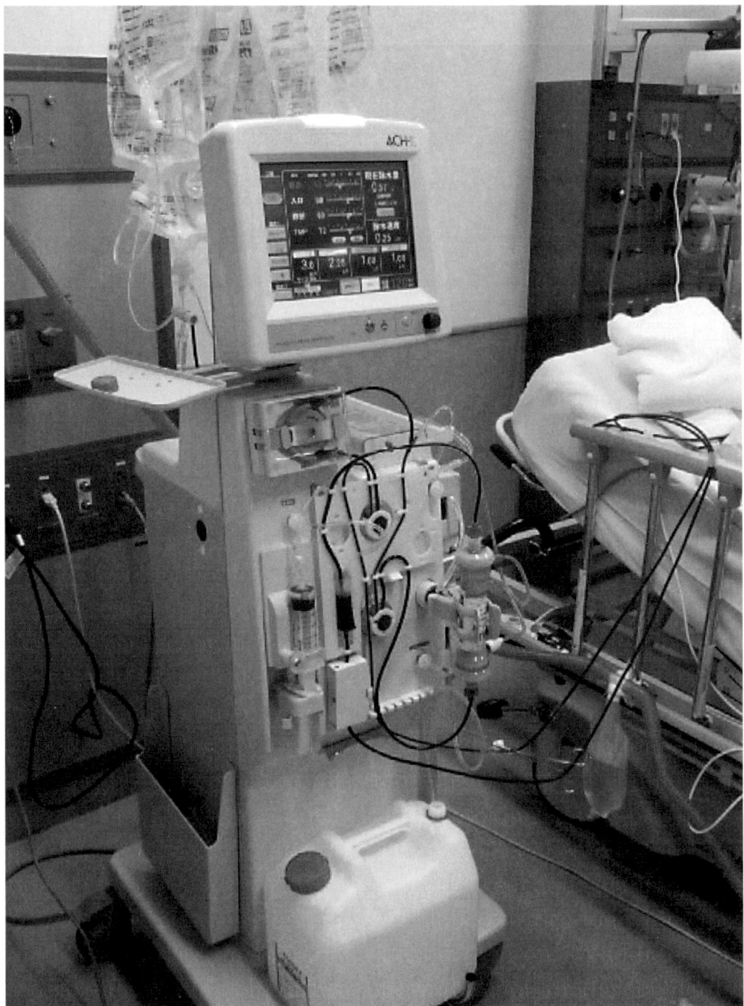

Fig. 1. CRRT console system.

automatically stop the procedure or reduce the blood flow rate within the variable pressure range (table 3).

Flow Rate of Dialysate, Substitution and Ultrafiltrate

The flow rate of each solution must be checked routinely; dialysate flow rate for continuous hemodialysis, ultrafiltrate flow rate for continuous hemofiltration, and both for continuous hemodiafiltration. It has recently been reported that increasing ultrafiltrate flow rate is helpful for improvement of prognosis [2, 3] so extreme precision is required on the roller pump system. Usually,

Table 2. Setting of CRRT for adult patients

Blood access	double lumen venous catheter internal jugular: 12 Fr external diameter, 13–15 cm length femoral: 12 Fr external diameter, 20–25 cm length
Membrane	continuous hemofilter, membrane surface 0.7–1.0 m^2
Bedside console	console system specific for CRRT
Blood circuit	disposable blood circuit kit specific for console
Anticoagulant	nafamostat mesilate 30 mg/h (or 0.5 mg/h/kg) Activated coagulation time monitoring 150–170 s
Solution	solution with sodium bicarbonate prescribed for CRRT
Blood flow rate	80–120 ml/min
Modality	continuous hemodialysis; continuous hemodialfiltration; continuous hemofiltration; continuous extracorporeal ultrafiltration
Dialysate flow rate	0.3–2.0 l/h
Replacement flow rate	0.3–2.0 l/h
Ultrafiltration rate	adequately set from volume balance and session time
Session time	6–24 h/day

a 10–20% increase in substitution flow rate for blood flow rate is recommended in continuous hemofiltration avoid the occlusion of the hemofilter, with a 30% increase in dialysate flow rate for blood flow rate in intermittent hemodialysis.

Fluid Flow Control System
Water balance is crucial for the prognosis of critical patients, and hyper- or hypovolemic status must be avoided. In this regard, the precision of the pump system has an important influence. On a CRRT console, fluid flow is controlled by flow rate using a roller pump, by volume change, and by weight monitoring. In our system, the difference between flow rate monitoring and other monitoring systems was within 1%. Systematic stability should be maintained in order to avoid a fatal error concerning the volume control [4].

Pump System for Anticoagulant
A syringe system is usually employed with the pump system for anticoagulation. Drugs for anticoagulation are injected continuously in any hemoperfusion

Table 3. Difficulties and responses in CRRT

Alarm indication	Main reason	Response
Inflow failure	kink in circuit (access side)	release kink
	occlusion of catheter	catheter change
	catheter tip attachment to vessel wall	decrease Qb, change body position
High filter pressure	occlusion of circuit (access side)	exchange filter and circuit, adjust Qb
	occlusion of hemofilter	exchange filter and circuit, adjust Qb
High venous pressure	occlusion of circuit (return side)	exchange circuit
	occlusion of catheter	change catheter
	kink of circuit (return side)	release kink
Low venous pressure	connection failure	confirm connection of circuit, catheter, filter
High transmembrane, low effluent pressure	occlusion of hemofilter	exchange filter and circuit, adjust Qb
High effluent pressure	kink in circuit (return side)	release kink
Air detected	air detected	remove air
	sensor off circuit	re-install sensor
Anticoagulant running out	anticoagulant running out	supply anticoagulant
Fluid running out	fluid running out	supply fluids
	air detected	remove air
	sensor off circuit	re-install sensor
Blood contamination	blood contamination	visually confirm, discontinue session
	sensor off circuit	re-install sensor
Electric power failure	socket off or code snapping	check power supply
	voltage down	check power supply and number of connected machines

Qb = blood flow rate.

therapy, and it is important that staff monitor closely the speed of injection and remaining quantities of the drugs. CRRT is often carried out for patients with bleeding complications, and nafamostat mesilate is the first choice anticoagulant in Japan. In our unit, activated coagulation time is routinely monitored when nafamostat mesilate is used.

Pressure Monitoring

Pressure monitoring is one of the most important check points in extracorporeal-circulation techniques that include CRRT. Thus, on a CRRT console, many pressure monitors are set and accompanied with alarms to ensure safety of the therapy.

Filter Pressure Monitor

The filter pressure monitor shows the pressure of the inflow side of the hemofilter, and it detects abnormal changes in blood flow rate, venous pressure and filtrate pressure. Increased filter pressure means occlusion of the filter fibers or venous drip chamber, in which case prompt exchange of the hemofilter or circuit set should be considered. Decreased filter pressure is usually the result of inflow failure, and the vascular access should be checked.

Return Pressure Monitor

The return pressure monitor shows the intrapressure of the venous drip chamber, the status of the catheter and blood flow. Increased return pressure is usually caused by occlusion of the venous drip chamber or the double lumen catheter.

Effluent Pressure Monitor

The effluent pressure monitor shows the intrapressure of the dialysate circulate system. The pressure depends on the setting of the fluid and blood flow control, venous pressure, and ultrafiltration rate. Decreased effluent pressure is usually the result of occlusion of the pore on the filter.

Transmembrane Pressure Monitor

A transmembrane pressure monitor shows the pressure differences between the blood side and dialysate side of filter fiber. The pressure is calculated as: transmembrane pressure = (filter pressure + return pressure)/2 – effluent pressure. Increased transmembrane pressure is caused by coagulative occlusion of the hemofilter, which should be exchanged quickly.

Blood Inflow Sensor

As a negative pressure sensor, a pillow-type small bag is installed in the blood circuit system. Attending staff should visually note the rough amount of blood inflow. When enough blood inflow is not provided, the bag will shrivel, and an electrical alarm will sound following the inflow pump pausing. A low level of blood inflow induces coagulative occlusion of the blood circuit.

Air Detector

Attenuation of ultrasonic propagation detects air bubbles in the blood circuit system. A common reason for air contamination is difficulties in the connection between blood access catheter and the blood circuit. Usually the detection of an air bubble stops all the pumps, and staff should check carefully the whole CRRT system.

Blood Detector in Filtrate Circuit

Blood contamination to the filtrate circuit is detected by changes in the permeated brightness. A common reason for blood contamination is fiber damage. Changing the hemofilter and blood circuit should be considered when blood contamination is detected.

Temperature

Blood flow rate, the air outside the circuit system, and the contact of blood to dialysate can change the temperature of the solution in the circuit system. To maintain patient temperature, the solution temperature is monitored and controlled by warming.

Coagulative Occlusion of Hemofilter and Blood Circuit

The filter membrane is composed of polysulfone, polyacrylnitrile, cellulosetriacetate and polymethylmethacrylate, and is designed to fit to the console system. These specifications have an effect on the conditions and setting of therapy. Thus, staff should know and understand the specificities of each console system, and adjust the proper condition to avoid coagulative occlusion of the system. A simply designed circuit system usually provides low incidence of coagulative occlusion.

Management of CRRT Console System

With developments and improvements of CRRT systems, the treatment possibilities they afford are increasing yearly. As it is impossible to manage and control the system personally, a staff team should share the management of CRRT. Safety is the most important point of the management. Information on the console system (numbers, types, uses and error notes) should be worked-up together and shared. It also helps to contact the provider to further development a new console.

References

1 Kramer P, Wigger W, Rieger J, et al: Arteriovenous hemofiltration: a new and simple method for treatment of overhydrated patients resistant to diuretics. Klin Wochenschr 1977;55:1121–1122.

2 Bellomo R, Baldwin I, Cole L, et al: Preliminary experience with high-volume hemofiltration in human septic shock. Kidney Int 1998;66:182–185.

3 Ronco C, Bellomo R, Homel P, et al: Effects of different doses in continuous veno-venous haemofiltration on outcomes of acute renal failure: a prospective randomized trial. Lancet 2000;356:26–30.

4 Saudan P, Niederberger M, De Seigneux S, et al: Adding a dialysis dose to continuous hemofiltration increases survival in patients with acute renal failure. Kidney Int 2006;79:1312–1317.

Yoshihisa Yamashita
School of Biomedical Engineering, Faculty of Health and Medical Care, Saitama Medical University
1397-1 Yamane, Hidaka
Saitama 350-1241 (Japan)
Tel./Fax +81 42984 4916, E-Mail yysmucet@saitama-med.ac.jp

Suzuki H, Hirasawa H (eds): Acute Blood Purification.
Contrib Nephrol. Basel, Karger, 2010, vol 166, pp 134–141

Cytokine Adsorbing Columns

Takumi Taniguchi

Intensive Care Unit, Kanazawa University Hospital, Kanazawa, Japan

Abstract

Sepsis induces the activation of complement and the release of inflammatory cytokines such as TNF-α and IL-1β. The inflammatory cytokines and nitric oxide induced by sepsis can decrease systemic vascular resistance, resulting in profound hypotension. The combination of hypotension and microvascular occlusion results in tissue ischemia and ultimately leads to multiple organ failure. Recently, several experimental and clinical studies have reported that treatment for adsorption of cytokines is beneficial during endotoxemia and sepsis. Therefore, the present article discusses cytokine adsorbing columns. These columns, such as CytoSorb, CYT-860-DHP, Lixelle, CTR-001 and MPCF-X, the structures of which vary significantly, have excellent adsorption rates for inflammatory cytokines such as TNF-α, IL-1β, IL-6 and IL8. Many studies have demonstrated that treatment with cytokine adsorbing columns has beneficial effects on the survival rate and inflammatory responses in animal septic models. Moreover, several cases have been reported in which treatment with cytokine adsorbing columns is very effective in hemodynamics and organ failures in critically ill patients. Although further investigations and clinical trials are needed, in the future treatment with cytokine adsorbing columns may play a major role in the treatment of hypercytokinemia such as multiple organ failure and acute respiratory distress syndrome.

Sepsis induces the activation of complement and the release of inflammatory cytokines such as TNF-α and IL-1β [1]. The inflammatory cytokines, particularly TNF-α, IL-1β and IL-6, can in turn trigger secondary inflammatory cascades, including the production of cytokines, lipid mediators and reactive oxygen species [2, 3]. The inflammatory cytokines and nitric oxide induced by sepsis can decrease systemic vascular resistance, resulting in profound hypotension [4, 5]. The combination of hypotension and microvascular occlusion results in tissue ischemia and ultimately leads to multiple organ failure [6].

Table 1. Structures, adsorption rates of cytokines in vitro, and animal study in cytokine adsorbing columns

	Column				
	CytoSorb	CYT-860-DHP	Lixelle	CTR-001	MPCF-X
Structure	polystyrene divinyl co-polymer beads	polystyrene-based conjugated fiber	porous cellulose beads	porous cellulose beads	cellulose beads
Adsorption rate					
Methods	in vitro circuit (1 h)	batchwise (2 h)	batchwise (2 h)	batchwise (2 h)	batchwise (1 h)
TNF-α	<50%	20%	31.2%	53%	100%
IL-1β		97%	98.5%	98%	
IL-6	<50%	92%	82.9%	80%	98.9%
IL-8		99%	99.9%	80%	70%
Animal study					
Aminal	rat		rat	rat	
Methods	endotoxin injection, cecal ligation and puncture		endotoxin injection	endotoxin injection	
Time	240 min		120 and 180 min	120 min	

Extracorporeal treatments for cleaning blood have been developed as systemic therapies to control refractory diseases. For the treatment of sepsis and septic shock, plasma exchange, hemoperfusion with polymyxin B immobilized fibers or filtration-and-dialysis treatments have been evaluated in septic patients. Several studies have reported improved hemodynamics and organ dysfunction, and inhibition of cytokine responses in patients with sepsis [7–9]. On the other hand, several reports have shown that these treatments do not improve the outcome or organ dysfunction and do not inhibit the elevation of cytokines in septic patients [10, 11]. Thus it is unclear whether these treatments are beneficial and remove cytokines in septic patients. Recently, several experimental and clinical studies have reported that the treatment for adsorption of cytokines is beneficial during endotoxemia and sepsis [12–26]. This article explains some cytokine absorbing columns (table 1).

CytoSorb Cartridge

Structure

CytoSorb (MedaSorb Technologies, Princeton, N.J., USA) [12–14] contains polystyrene divinyl benzene copolymer beads with a biocompatible polyvinylpyrrolidone coating. Each bead is 300–800 μm in size and each gram of material has a surface area of 850 m^2.

Cytokine and Toxin Adsorption in vitro

An in vitro circuit and cytokine-rich blood taken from endotoxin-challenged rats were used to determine the rate of removal of TNF-α, IL-6 and IL-10 [12]. TNF-α, IL-6 and IL-10 were rapidly removed with <50% of the initial concentrations present after 1 h of circulation through the CytoSorb cartridge.

Treatment in vivo

Several studies have evaluated the effects of treatment with CytoSorb cartridges on mortality and inflammatory responses to endotoxin-induced shock in rats [12, 13]. These studies observed that treatment with CytoSorb cartridges, carried out for 240 min after endotoxin injection, significantly reduced the high mortality rate and inhibited inflammatory responses in endotoxin-induced shock in rats. Moreover, the CytoSorb treatment reduced cytokines, improved mean arterial pressure and resulted in better short-term survival in cecal ligation and puncture-induced septic shock in rats.

Treatment in Patients

The CytoSorb treatment removed inflammatory cytokines such as TNF-α and IL-6, but not IL-10, in 8 brain-dead subjects [14]. Moreover, this study observed no adverse effects of the treatment. These findings suggest that the CytoSorb treatment may have beneficial effects in patients with hypercytokinemia.

CYT-860-DHP

Structure

The adsorbent used to construct the CYT-860 (Toray Industries Inc., Tokyo, Japan) [15] is prepared by chemical modification of a polystyrene-based conjugated fiber reinforced with polypropylene and originally developed as an exotoxin adsorber. Polystyrene is used as a fiber filter material packed in a polymyxin B-immobilized hemoperfusion cartridge for endotoxin removal. 4-(4-{12-[N'-(4-chlorophenyl)ureylene]-1,4,7,10-tetraazadodecylene}-acetylacetoamido) groups are covalently immobilized on the polystyrene component of the adsorbent fiber as adsorption ligands.

Cytokine and Toxin Adsorption in vitro
The adsorption ability of various molecules, such as cytokines, superantigens, and other proteins was examined by in vitro batchwise adsorption, incubating for 120 min. The adsorption rates were 20% for TNF-α, 97% for IL-1β, 92% for IL-6, 99% for IL-8, 83% for IL-10 and 96% for TSST-1, respectively.

Treatment in vivo
Effects of treatment with the CYT-860-DHP in septic shock models have not yet been evaluated in detail.

Treatment in Patients
The CYT-860-DHP treatment reduced blood cytokine concentrations and improved the general condition in 7 critically ill patients with hypercytokinemia [15]. This finding suggests that the CYT-860-DHP treatment may have beneficial effects in patients with hypercytokinemia.

Lixelle Column

Structure
Lixelle (Kanaka Corp., Osaka, Japan) [16–20] selectively adsorbs β2-microglobulin for the treatment of dialysis-related amyloidosis in Japan. Lixelle is composed of porous cellulose beads to which a hydrophobic organic compound with a hexadecyl alkyl chain has been covalently bound to the surface as a ligand.

Cytokine and Toxin Adsorption in vitro
The adsorption ability of various molecules, such as cytokines, was examined by in vitro batchwise adsorption, incubating for 120 min. The adsorption rates were 31.2% for TNF-α, 98.5% for IL-1β, 98.0% in IL-1Ra, 82.9% for IL-6, and 99.9% for IL-8, respectively [17].

Treatment in vivo
Several studies have evaluated the effects of treatment with the Lixelle column on the mortality and inflammatory responses to endotoxin-induced shock in rats [18, 19]. These studies observed that treatment with the Lixelle column, which has been carried out for 120 or 180 mins after endotoxin injection, drastically reduced the high mortality rate and inhibited cytokine responses in endotoxin-induced shock in rats.

Treatment in Patients
The Lixelle treatment increased blood pressure and 5 critical patients recovered from shock status. Moreover, the Lixelle treatment for 4 h was

observed to remove the inflammatory cytokines [20]. These findings suggest that the Lixelle treatment may have beneficial effects in patients with hypercytokinemia.

CTR-001 Column

Structure

CTR-001 [21–25], developed by Kaneka Corp., was based on a predecessor, Lixelle. CTR-001 is composed of porous cellulose beads to which a hydrophobic organic compound with a hexadecyl alkyl chain has been covalently bound to the surface as a ligand, the same as Lixelle. The CTR-001 beads are spherical with a diameter of approximately 460 μm and contain about 30 μmol of immobilized ligand per gram of suction-dried beads. When visualized using scanning electron microscopy, the adsorbent was found to be highly porous.

Cytokine and Toxin Adsorption in vitro

The serum concentrations of recombinant cytokines decreased remarkably in the batch adsorption test utilizing CTR-001. The adsorption rates of these proteins were 53% for TNF-α, 98% for IL-1β, 80% for IL-6, 80% for IL-8, 96% for IL-4 and 84% for IL-10, respectively. The serum concentrations of toxins also decreased remarkably. The adsorption rates of these toxins were 66% for SEA, 70% for SEB, 55% for SEC and 89% for TSST-1, respectively.

Treatment in vivo

Several studies have evaluated the effects of treatment with the CTR-001 column on mortality and inflammatory responses to endotoxin-induced shock in rats [21–24]. These studies observed that treatment with the CTR-001 column, which has been carried out for 120 mins after endotoxin injection, drastically and dose-dependently reduced the high mortality rate and inhibited inflammatory responses in endotoxin-induced shock in rats. Moreover, the CTR treatments carried out for 120 min after the endotoxin injection, markedly reduced the high mortality rate and inhibited inflammatory cytokine responses in endotoxin-exposed rats, the same as the endotoxin absorbing column (PMX column) treatment.

Treatment in Patients

The CTR treatment was effective for reducing the inflammatory cytokines such as TNF-α and IL-6 during severe acute pancreatitis in a 46-year-old male [25].

MPCF-X

Structure
MPCF-X [26] is modified by coating the surface of the adsorbent in CF-X with 2-methacryloxyethyl phosphorylcholine. CF-X consists of cellulose beads cross-linked with hexamethylene-di-isocyanate and was developed for the treatment of immunologic diseases such as systemic lupus erythematosus and rheumatoid arthritis.

Cytokine and Toxin Adsorption in vitro
The adsorption ability of various molecules, such as cytokines, and other proteins was examined by in vitro batchwise adsorption, incubating for 60 min. The adsorption rates were 100% for TNF-α, 98.9% for IL-6, 70% for IL-8 and 88.7% for IL-10, respectively.

Treatment in vivo and in Patients
Effects of treatment with the MPCF-X column have not yet been evaluated in detail in septic shock models or in critical patients.

Discussion

This chapter discusses the structure, cytokine adsorption rates and effects of treatment in vivo and in patients for several cytokine adsorbing columns. All columns in the present article, which structures are quite different, have excellent adsorption rates of inflammatory cytokines such as TNF-α, IL-1β, IL-6 and IL8. Many studies demonstrated that treatment with cytokine adsorbing columns had beneficial effects on the survival rate and inflammatory responses in animal septic models. Moreover, several cases have been reported in which treatment with cytokine adsorbing columns was very effective on hemodynamics and organ failure in critically ill patients.

Thus, this article indicates that the control of inflammatory cytokines cytokine adsorption may improve hemodynamics and organ dysfunction in septic or septic shock patients. However, some patients with severe septic shock may see no beneficial effects on hemodynamics and organ failure, as is the case with plasma exchange, hemoperfusion with polymyxin B immobilized fibers or filtration-and-dialysis treatments. It is very difficult to judge whether treatment with cytokine adsorbing columns should be performed in severe septic patients. Further investigations and clinical trials are needed.

We hope that treatment with cytokine adsorbing columns may play a major role in the treatment of hypercytokinemia such as multiple organ failure and acute respiratory distress syndrome in the future.

References

1 Cohen J: The immunopathogenesis of sepsis. Nature 2000;420:885–891.
2 Kuhns DB, Alvord WG, Gallin JI: Increased circulating cytokines, cytokine antagonists, and E-selectin after intravenous administration of endotoxin in human. J Infect Dis 1995;171:145–152.
3 Calandra T, Echtenacher B, Roy DL, Pugin J, Metz CN, Hultner L, Heumann D, Mannel D, Bucala R, Glauser MP: Protection from septic shock by neutralization of macrophage migration inhibitory factor. Nat Med 2000;6:164–170.
4 Torre-Amione G, Kapadia S, Benedict C, Oral H, Young JB, Mann DL: Proinflammatory cytokine levels in patients with depressed left ventricular ejection fraction: a report from the Studies of Left Ventricular Dysfunction (SOLVD). J Am Coll Cardiol 1996;27:1201–1206.
5 Carlson DL, Willis MS, White DJ, Horton JW, Giroir BP: Tumor necrosis factor-alpha-induced caspase activation mediates endotoxin-related cardiac dysfunction. Crit Care Med 2005;33:1021–1028.
6 Natanson C, Eichenholz PW, Danner RL, Eichacker PQ, Hoffman WD, Kuo GC, Banks SM, MacVittie TJ, Parrillo JE: Endotoxin and tumor necrosis factor challenges in dogs simulate the cardiovascular profile of human septic shock. J Exp Med 1989;169:823–832.
7 Bellomo R, Tipping P, Boyce N: Continuous veno-venous hemofiltration with dialysis removes cytokines from the circulation of septic patients. Crit Care Med 1993;21:522–526.
8 Stegmayr BG: Plasmapheresis in severe sepsis or septic shock. Blood Purif 1996;14:94–101.
9 Bengsch S, Boos KS, Nagel D, Seidel D, Inthorn D: Extracorporeal plasma treatment for the removal of endotoxin in patients with sepsis: clinical results of a pilot study. Shock 2005;23:494–500.
10 Reinhart K, Meier-Hellmann A, Beale R, Forst H, Boehm D, Willatts S, Rothe KF, Adolph M, Hoffmann JE, Boehme M, Bredle DL; EASy-Study Group: Open randomized phase II trial of an extracorporeal endotoxin adsorber in suspected Gram-negative sepsis. Crit Care Med 2004;32:1662–1668.
11 Vincent JL, Laterre PF, Cohen J, BurchardiH, Bruining H, Lerma FA, Wittebole X, De Backer D, Brett S, Marzo D, Nakamura H, John S: A pilot-controlled study of a polymyxin B-immobilized hemoperfusion cartridge in patients with severe sepsis secondary to intra-abdominal infection. Shock 2005;23:400–405.
12 Kellum JA, Song M, Venkataraman R: Hemoadsorption removes tumor necrosis factor, interleukin-6, and interleukin-10, reduces nuclear factor-κB DNA binding, and improves short-term survival in lethal endotoxemia. Crit Care Med 2004;32:801–805.
13 Peng ZY, Carter MJ, Kellum JA: Effects of hemoadsorption on cytokine removal and short-term survival in septic rats. Crit Care Med 2008;36:1573–1577.
14 Kellum JA, Venkataraman R, Powner D, Elder M, Hergenroeder G, Carter M: Feasibility study of cytokine removal by hemoadsorption in brain-dead humans. Crit Care Med 2008;36:268–272.
15 Kobe Y, Oda S, Matsuda K, Nakamura M, Hirasawa H: Direct hemoperfusion with a cytokine-adsorbing device for the treatment of persistent or severe hypercytokinemia: a pilot study. Blood Rurif 2007;25:446–453.
16 Tsuchida K, Takemoto Y, Nakamura T, Fu O, Okada C, Yamagami S, Kishimoto T: Lixelle adsorbent to remove inflammatory cytokines. Artif Organs 1998;22:1064–1067.
17 Tsuchida K, Yoshimura R, Nakatani T, Takemoto Y: Blood purification for critical illness: cytokines adsorption therapy. Ther Apher 2006;10:25–31.
18 Nakatani T, Tsuchida K, Fu O, Sugimura K, Takemoto Y: Effects of direct hemoperfusion with a beta2-microglobulin adsorption column on hypercytokinemia in rats. Blood Purif 2003;21:145–151.
19 Tsuda K, Taniguchi T: Effects of extracorporeal treatment with Lixelle on the mortality and inflammatory responses to endotoxin-induced shock in rats. Ther Apher 2006;10:49–53.
20 Tsuchida K, Takemoto Y, Sugimura K, Yoshimura R, Nakatani T: Direct hemoperfusion by using Lixelle column for the treatment of systemic inflammatory response syndrome. Int J Mol Med 2002;10:485–488.

21 Taniguchi T, Hirai F, Takemoto Y, Tsuda K, Yamamoto K, Inaba H, Sakurai H, Furuyoshi S, Tani N: A novel adsorbent of circulating bacterial toxins and cytokines; the effect of direct hemoperfusion with CTR column for the treatment of experimental endotoxemia. Crit Care Med 2006;34:800–806.

22 Taniguchi T, Takemoto Y, Tsuda K, Inaba H, Yamamoto K: Effects of posttreatment with direct hemoperfusion using a CTR column on mortality and inflammatory responses to endotoxin-induced shock in rats. Blood Purif 2006;24:460–464.

23 Taniguchi T, Kurita A, Mukawa C, Yamamoto K, Inaba H: Dose-related effects of direct hemoperfusion using a cytokine adsorbent column for the treatment of experimental endotoxemia. Intensive Care Med 2007;33:539–544.

24 Taniguchi T, Kurita A, Yamamoto K, Inaba H: Comparison of a cytokine adsorbing column and an endotoxin absorbing column for the treatment of experimental endotoxemia. Transfus Apher Sci 2009;40:55–59.

25 Saotome T, Endo Y, Sasaki T, Tabata T, Hamamoto T, Fujino K, Eguchi Y, Tani T, Fujiyama Y: A case of severe acute pancreatitis treated with CTR-001 direct hemoperfusion for cytokine apheresis. Ther Apher 2005;9:367–371.

26 Oda S, Hirasawa H, Matsuda K, Nakanishi K, Matsuda K, Nakamura M, Ikeda H, Sakai M: Cytokine adsorptive property of various adsorbents in immunoadsorption columns and a newly developed adsorbent. Blood Purif 2004;22:530–536.

Takumi Taniguchi
Intensive Care Unit, Kanazawa University Hospital
13-1 Takara-machi
Kanazawa 920-8641 (Japan)
Tel. +81 76 265 2826, Fax +81 76 234 0973, E-Mail ttaniyan@med.kanazawa-u.ac.jp

Suzuki H, Hirasawa H (eds): Acute Blood Purification.
Contrib Nephrol. Basel, Karger, 2010, vol 166, pp 142–149

Plasma Dia-Filtration for Severe Sepsis

Yutaka Eguchi

Department of Critical and Intensive Care Medicine, Shiga University of Medical Science, Otsu City, Japan

Abstract

The mortality rate in severe sepsis is 30–50%, and independent liver and renal dysfunction impacts significantly on hospital and intensive care mortality. If 4 or more organs fail, mortality is >90%. Recently, we reported a novel plasmapheresis – plasma dia-filtration (PDF) – the concept of which is plasma filtration with dialysis. PDF employs a plasma separator that has a sieving coefficient of 0.3 for albumin and which requires flowing dialysate outside the hollow fiber. For substitute liquid, 1,200 ml of fresh frozen plasma followed by 50 ml of 25% albumin solution is used for 8 h as 1 session. In a single-center study, 24 patients with septic shock were admitted to the ICU, then 37.7 ± 30.0 h later, 7 patients received PDF. The patients' Sequential Organ Failure Assessment (SOFA) scores had increased from 14.9 ± 3.6 on ICU admission to 17.1 ± 3.0 before PDF procedure. PDF was performed, with an average of 7.4 ± 4.4 sessions (range 3–15) per patient. Five patients survived after day 28, thus the 28-day mortality rate was 29%. In our multicenter study, 33 patients with severe sepsis who simultaneously suffered from liver dysfunction were enrolled and received PDF. On average, 12.0 ± 16.4 sessions (range 2–70) per patient were performed. The 28-day mortality rate was 36.4%, while the predicted death rate was 68.0 ± 17.7%. These findings suggest that PDF is a simple modality and may become a useful strategy for treatment of patients with septic multiple organ failure.

The mortality of severe sepsis is 39.2% (22–56.8%) [1], which has been unacceptably high. Patients with severe sepsis sometimes have not withdrawn from septic shock and/or multiple organ failure (MOF) [2]. It has been frequently reported that coexisting and indeed independent liver and renal dysfunction impact significantly on hospital and intensive care mortality [3], when present as part of MOF. Mortality is >90% if 4 or more organs fail [4].

Uremic toxin bound on albumin interacts with the free levels of many other substances in serum. Albumin is a universal carrier, transporting numerous chemically diverse endogenous and exogenous molecules and affecting their distribution and metabolism. The accumulation of albumin-bound toxin (ABT) results in impaired protein binding of other compounds because of competitive ligand binding. This impairs protein anionic drugs and endogenous molecules such as bilirubin (Bil). Albumin dialysis is the most commonly used artificial liver support system and provides effective elimination of ABT [5].

Clinical and molecular biology research has suggested that cytokines and other septic mediators, such as endotoxin, histones [6] and high morbility group box-1 (HMGB-1) [7], contribute to the pathogenesis of development into MOF. The removal of these substances may be considered to reduce the mortality. PMX direct hemoperfusion [8] and high flow/volume continuous hemodiafiltration (CHDF) using PMMA membrane [9] have been widely used, especially in Japan, for elimination of endotoxin and cytokines. Recently, high-cutoff-point membranes with moderately increased pore size were developed and conducted to a phase 1 study in sepsis complicated by acute renal failure [10].

We have contrived a novel plasmapheresis technique, called plasma dia-filtration (PDF), and reported that Bil was removed effectively with treatment for acute hepatic insufficiency [11]. PDF seems to eliminate the low to medium molecular weight substances, including ABT, cytokines, histones and HMGB-1. Therefore, PDF may be considered to reduce the mortality of severe sepsis by preventing MOF. In this report, we evaluate the effect of PDF on clinical parameters in patients for the treatment of septic MOF with liver dysfunction.

Patients and Methods

Study Population

The study was approved by the human research ethics committee of Shiga University of Medical Science. Written informed consent was obtained from each patient's next of kin. Inclusion criteria comprised the simultaneous presence of consensus criteria for sepsis [12].

We calculated Acute Physiology and Chronic Health Evaluation II (APACHE II) [13] and Sequential Organ Failure Assessment (SOFA) [14] scores on both ICU admission and the day before PDF procedure. The organ failure was defined by organ dysfunction criterion (table 1) [15]. The liver dysfunction criteria were defined as the following (at least 2 items): serum total Bil ≧7 mg/dl, hepaplastin tent ≦40% and level of consciousness loss greater than Coma Grade II. The predictive mortality was estimated by APACHE II and we also determined the number and combination of MOF [2].

Patients

Septic Shock in a Single-Center Study. All patients were screened for a diagnosis of septic shock in spite of sufficient fluid resuscitation on ICU admission in Shiga University of Medical Science Hospital. Twenty-four patients were eligible and enrolled in this retro-

Table 1. Criteria for dysfunction organs or systems (patients had to meet at least 1 criterion) [15]

Cardiovascular system dysfunction
Arterial systolic blood pressure ≦90 mmHg or mean arterial pressure ≦70 mmHg for at least 1 h despite adequate fluid resuscitation, adequate intravascular volume status or the use of vasopressors in an attempt to maintain blood and arterial pressures at values greater than these.
Kidney dysfunction
Urine output <0.5 ml/kg of body weight/h for 1 h, despite adequate fluid resuscitation.
Respiratory system dysfunction
Pao_2 to Fio_2 ratio ≦250 in the presence of other dysfunctional organs or systems, or ≦200 if lung is only dysfunctional organ.
Hematologic dysfunction
Platelet count <80,000/mm^3 or decreased by 50% in the 3 days preceding enrollment.
Unexplained metabolic acidosis
pH ≦7.30 or base deficit ≧5.0 mmol/l in association with plasma lactate level >1.5 times upper limit of normal for the reporting laboratory.

spective analysis between March 2006 and February 2008. Seven patients developed MOF with liver dysfunction, and then received PDF. PDF was performed for 8 h once a day as 1 session or 24 h continuously as 3 sessions, depending on the patient's condition.

Severe Sepsis in a Multicenter Study. We screened for a diagnosis of severe sepsis all patients who suffered from liver dysfunction in Shiga University of Medical Science Hospital, Nishi-Kyoto Hospital, Hikone-City Hospital and Akita University Hospital. Thirty-three patients were eligible and enrolled in this retrospective study between October 2005 and March 2009. PDF was performed for 8 h once a day as 1 session or 24 h continuously as 3 sessions, according to the patient's condition.

PDF Technique

PDF employed the ethylene-vinyl alcohol copolymer 1.0 m^2 membrane plasma separator 'Evacure' EC-2A (Kawasumi Chemical Inc., Tokyo, Japan), which has a sieving coefficient of 0.3 for albumin. PDF requires flowing dialysate outside the Evacure hollow fiber (fig. 1), which is expected to improve the removal efficiency of medium molecular weight substances. The flow rate of the blood, dialysate, substitute and additional one were 80–100 ml/min, 600 ml/h, 0–450 ml/h according to the rate of water-elimination and 150 ml/h, respectively. The filtration rate was 600 ml/h. We used Sublood BS (Fuso Pharmaceutical Industries Ltd., Osaka, Japan) as dialysate and substitute.We added as substitute from the additional fluid line 15 units (1,200 ml) of fresh frozen plasma, followed by 50 ml of 25% albumin, considering the loss of albumin by diffusion. As an

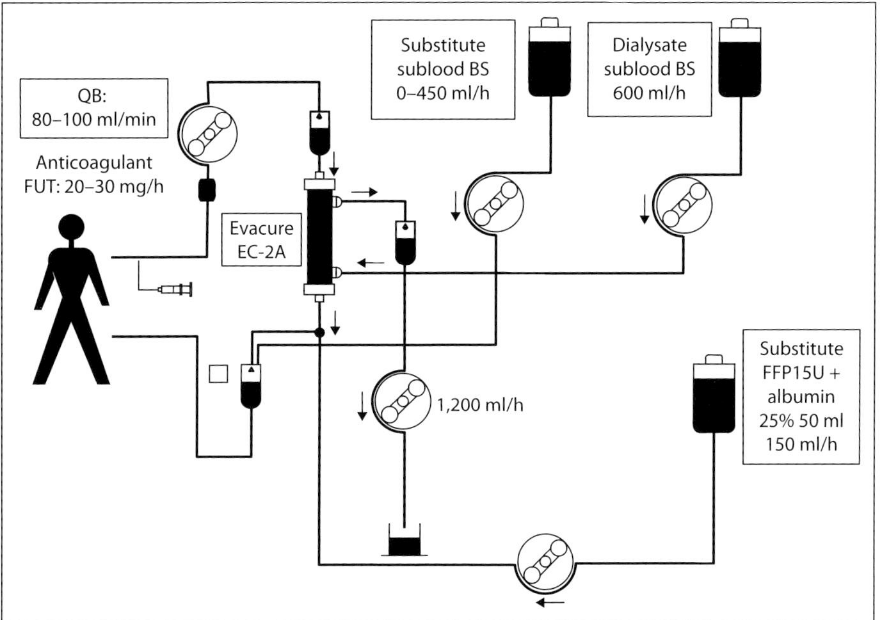

Fig. 1. Schematic diagram of the PDF. FUT = Nafamostat mesilate.

anticoagulant, nafamostat mesilate (Torii Pharmaceutical Co. Ltd., Tokyo, Japan) was used at a rate of 20–35 mg/h.

Statistical Analysis

Descriptive data are expressed as means ± SD.

Results

PDF for Septic Shock in a Single Center Study

Twenty-four patients were admitted to the ICU for septic shock and 7 patients received PDF. Of these patients, 3 pairs received PDF only, PDF following CHDF and PDF following PMX-DHP, respectively, and 1 patient received PDF following CHDF plus PMX-DHP. A total of 52 sessions were performed (average of 7.4 ± 4.4 per patient, range 3–15). The time when PDF was performed was 37.7 ± 30.7 h (range 5.5–92 h) after ICU admission. Before the PDF procedure, the clinical characteristics of the patients who received PDF were compared with those on ICU admission (table 2). The SOFA score increased before PDF procedure, compared with that on ICU admission. Two patients who received PDF died, one 3 days after PDF procedure (from hypotension), and the other 14 days after (from respiratory failure). The 28-day survival rate was 71%, whereas the predicted mortality rate was 74.4 ± 8.6%.

Table 2. Clinical characteristics of the patients in a single-center study

	Overall	Patients with PDF	
	on ICU admission	on ICU admission	before PDF procedure
Cases, n	24	7	
Sex (male/female), n	12/12	5/2	
Age, years	64.5±15.0	59.4±18.1	
28-day mortality, n (%)	5/24 (21)	2/7 (29)	
APACHE II score	30.4±7.8	33.4±4.4	31.1±5.2
Predicted mortality, %	72.5±22.6	86.5±9.2	80.0±12.8
SOFA score	12.8±3.4	14.9±3.6	17.1±3.0
Failed organs, n	3.6±0.9	4.1±0.4	3.1±1.4
Predicted mortality, %	65.9±18.0	74.4±8.6	57.0±23.2

Table 3. Clinical characteristics of the patients in a multicenter study

	Overall	Survivors	Non-survivors
Case, n	33	21	12 (36.4%)
Age, years	61.5±18.5	58.9±21.4	66±11.4
Sex (male/female), n	24/9	15/6	9/3 (37.5/33.3%)
Sessions, n	12.0±16.4 (range 2–70)	15.9±19.5 (range 2–70)	5.2±3.2 (range 1–11)
Failed organs, n	3.7±1.0	3.3±1.0	4.3±0.7
Number of patients with			
2 failed organs	5	5	0 (0%)
3 failed organs	8	7	1 (12.5%)
4 failed organs	13	7	6 (46.2%)
5 failed organs	7	2	5 (71.4%)
Predicted death rate	68.0±17.7%	60.5±18.1%	81.0±4.8%

Patients with Severe Sepsis with Liver Dysfunction in a Multicenter Study
The clinical characteristics of the patients are given in table 3. PDF treatment was given to 33 patients with severe sepsis who simultaneously suffered from liver dysfunction. Organ failure consists of the following 5 criteria: cardiovascular system, kidney, respiratory system, hematological and metabolic dysfunctions (except liver dysfunction; table 1). The actual number of organ failures was 3.7 ±1.0 (plus 1 including liver dysfunction). Five patients had 2 organ failures, 8 had 3 failures, 13 patients had 4, and 7 had 5. Overall the 28-day survival rate was 63.6%. The 28-day mortality rate was better than that predicted. The former was 36.4%, and the latter was 68.0 ± 17.7%, which coincided with the previous report [2]. Among the patients with 2 organ failures, none died. The mortality of patients with 3 failed organs was 12.5%, for those with 4 it was 46.2%, and for 5 organ failures it was 71.4%. For the patients with over 4 organ failures (5 including liver dysfunction), it was difficult to survive after 28 days.

Discussion

Acute blood purification may eliminate pathological substances, such as ABT (albumin molecular weight 68 kDa), cytokine (8–54 kDa), histones (H1: 22 kDa, H2A and B: 13.7 kDa, H3: 15.7 kDa, H4: 112.2 kDa) and HMGB-1 (30 kDa), and to maintain the physiological substances that act as host defense, such as coagulation factors, hepatocyte growth factor and immunoglobulin. According to our unpublished observations, after 1 h of the PDF procedure (Qf 10 ml/min, Qd 10 ml/min) clearances (ml/min) were: BUN 18.4, creatinine 21.5, IL-6 18.1, IL-8 18.8 and total Bil 5.6. This indicates that PDF may be able to eliminate water-soluble toxins and cytokines.

Non-selective removal of such mediators with hemofiltration using standard membranes and filtration rates has had a limited effect on circulating cytokine levels and clinical outcomes. Recently, PMMA-CHDF was shown to increase urinary output in septic acute renal failure patients, and dramatically improve the survival rate in septic ARF patients [9]. Therefore, it is possible to assume that the cytokine is modulated by PMMA-CHDF in the treatment of septic ARF prior to development of MOF.

Recently, high-cutoff-point membranes have been introduced to remove medium molecular weight mediators by increasing the pore size of renal replacement membranes. Preliminary clinical studies show the high-cutoff-point membranes decrease plasma cytokine levels and the need for vasopressor therapy, with no reports of serious adverse effects [16]. However, reduction in serum albumin and lower small solute clearance was reported [10], therefore further investigation will be required. PDF is considered to be selective plasma filtration with dialysis, and therefore removes water-soluble substances and cytokines including ABT.

Albumin dialysis, such as MARS and Prometheus, is the most commonly used artificial liver support system and provides elimination of ABT and the water-soluble toxins that accumulate in liver failure. Cytokines play an important role in the pathogenesis of MOF in both sepsis and severe liver insufficiency. MARS and Prometheus cleared cytokines from plasma, but neither system is able to change serum cytokine levels [17–19], because the rate of cytokine production in patients with liver failure is higher than that of its elimination by MARS and Prometheus. These findings suggest that MARS and Prometheus cannot be useful in severe sepsis with liver failure.

More recently, histones killed endothelial cells in vitro and mimicked the symptoms of sepsis including organ failure, when injected into mice [6]. This observation gives rise to a possibility that elimination of histones may have to be developed and could have great promise in preventing death during sepsis. It was reported that patients with systematic inflammatory response syndrome received blood purification procedures such as continuous renal replacement therapy, endotoxin adsorption, and/or plasma exchange and 50% of the patients died because of with their extremely high SOFA score (≥18 or a ΔSOFA score ≥3) [20]. This finding suggests that such blood purification procedures limit the elimination of histones. In our single-center analysis, all of 4 patients with SOFA ≥18 and/or with ΔSOFA scores ≥3 survived 28 days after PDF procedure, which suggests that PDF may be able to eliminate histones in some part.

In conclusion, PDF is a simple modality and appears to eliminate water-soluble toxins and cytokines and, in some part, may therefore be useful in reducing mortality in severe sepsis by preventing MOF, especially with liver dysfunction.

Acknowledgement

The author wishes to acknowledge the contributions of T. Yoshioka, MD (Nishi Kyoto Hospital), H. Nakae, MD (Akita University Hospital) and Y. Kishi, MD (Hikone City Hospital).

References

1 Beale R, Reinhart K, Brunkhorst FM, Dobb G, Levy M, Martin G, Martin C, Ramsey G, Silva E, Vallent B, Vincent JL, Janes JM, Sarwat S, Williams MD, for the PROGRESS Advisory Board: Promoting Global Research Excellence in Severe Sepsis (PROGRESS): Lessons from an international sepsis registry. Infection 2009;37:222–232.

2 Padkin A, Goldfrad C, Brady AR, Young D, Black N, Rowan K: Epidemiology of severe sepsis occurring in the first 24 hrs in intensive care units in England, Wales, and Northern Ireland. Crit Care Med 2003;31:2332–2338.

3 Hoste EAJ, Clermont G, Kersten A, Venkataraman R, Angus DC, De Bacquer D, Kellum JA: RIFLE criteria for acute kidney injury are associated with hospital mortality in critically ill patients: a cohort analysis. Critical Care 2006;10:R73.

4 Abosaif NY, Tolba YA, Heap M, Russell J, El Nahas AM: The outcome of acute renal failure in the intensive care unit according to RIFLE: model application, sensitivity, and predictability. Am J Kidney Dis 2005;46: 1038–1048.

5 Williams R: Correction of disturbed pathophysiology of hepatic failure by albumin dialysis. Hepatobiliary Pancreat Dis Int 2008;7:19–24.

6 Xu J, Zhang X, Pelayo R, Monestier M, Ammollo CT, Semeraro F, Taylor FB, Esmon NL, Lupu F, Esmon CT: Extracellular histones are major mediators of death in sepsis. Nature Med 2009;15:1318–1321.

7 Wang H, Bloom O, Zhang M, Vishnubhakat JM, Ombrellino M, Che J, Frazier A, Yang H, Ivanova S, Borovikova L, Manogue KR, Faist E, Abraham E, Andersson J, Andersson U, Molina PE, Abumrad NN, Sama A, Tracey KJ: HMGB-1 as a late mediator of endotoxin lethality in mice. Science 1999;285:248–251.

8 Cruz DN, Antonelli M, Fumagalli R, Foltran F, Brienza N, Donati A, Malcangi V, Petrini F, Volta G, Pallavicini FMB, Rottoli F, Giunta F, Ronco C: Early use of polymyxin B hemoperfusion in abdominal septic shock. JAMA 2009;301:2445–2452.

9 Matsuda K, Moriguchi T, Harii N, Goto J: Comparison of efficacy between continuous hemodiafiltration with a PMMA membrane hemofilter and a PAN membrane hemofilter in the treatment of a patient with septic acute renal failure. Transfus Apher Sci 2009;40: 49–53.

10 Lee D, Haase M, Haase-Fielitz A, Paizis K, Goehl H, Bellomo R: A pilot, randomized, double-blind, cross-over study of high cut-off versus high-flux dialysis membranes. Blood Purif 2009;28:365–372.

11 Mori T, Eguchi Y, Shimizu T, Endo Y, et al: A case of acute hepatic insufficiency treated with novel plasmapheresis plasma dia-filtration for bridge use until liver transplantation. Therapeutic Apheresis 2002;6:463–466.

12 Members of Society of Critical Care Med Consensus Conference Committee: Definitions for sepsis and organ failure and guidelines for the use of innovative therapies in sepsis. Crit Care Med 1992;20:864–874.

13 Knaus WA, Draper EA, Wagner DP, et al: APACHE II: A severity of disease classification system. Crit Care Med 1985;13:818–829.

14 Vincent JL, Moreno R, Takala J, et al: The SOFA (Sequential Organ Failure Assessment) score to describe organ dysfunction/failure. On behalf of the working Group on Sepsis-Related Problems of the European Society of Intensive Care Medicine. Intensive Care Med 1996;22:707–710.

15 Bernard, GR, Vincent J-L, Laterre P-F, Larosa SP, Dhainut J-F, Lopez-Rodriguez A, Steingrub JS, Garber GE: Efficacy and safety of recombinant human activated protein C for severe sepsis. N Engl J Med 2001; 344:699–709.

16 Haase M, Bellomo R, Morger S, Baldwin I, Boyce N: High cut-off point membranes in septic acute renal failure: a systematic review. Int J Artif Organs 2007;30:1031–1041.

17 Stadlbanuer V, Krisper P, Aigner R, Haditsch B, Jung A, Lackner C, Stauber RE: Effect of extracorporeal liver support by MARS and Prometheus on serum cytokines in acute-on-chronic liver failure. Crit Care 2006;10:R169.

18 Llonen L, Koivusalo A-M, Hockerstedt K, Isoniemi H: Albumin dialysis has no clear effect on cytokine levels in patients with life-threatening liver insufficiency. Transplant Proc 2006;38:3540–3543.

19 Llonen L, Koivusalo A-M, Repo H, Hockerstedt K, Isoniemi H: Cytokine profiles in acute liver failure treated with albumin dialysis. Artif Organs 2008;32:52–60.

20 Kikuchi H, Maruyama H, Omori S, Kazama JJ, Gejyo F: The Sequential Organ Failure Assessment Score as a useful predictor for estimating the prognosis of systemic inflammatory response syndrome patients being treated with extracorporeal blood purification. Ther Apher Dial 2003;7:456–460.

Yutaka Eguchi, MD
Department of Critical and Intensive Care Medicine, Shiga University of Medical Science
520-2192 Seta Tsukinowa-cho
Otsu City, Shiga (Japan)
Tel. +81 77 548 2929, Fax +81 77 548 2929, E-Mail eguchi@belle.shiga-med.ac.jp

Suzuki H, Hirasawa H (eds): Acute Blood Purification.
Contrib Nephrol. Basel, Karger, 2010, vol 166, pp 150–157

Application of Polymyxin B Convalently Immobilized Fiber in Patients with Septic Shock

Hiromichi Suzuki[a] · Hisataka Shoji[b]

[a]Department of Nephrology, Saitama Medical University, Saitama, [b]Toray Medical Co., Tokyo, Japan

Abstract

Sepsis and septic shock are major causes of morbidity and mortality in the intensive care unit. Endotoxin produced by Gram-negative bacteria contributes to the pathogenesis of sepsis and septic shock. As an adsorbent, a polymyxin B convalently immobilized fiber (PMX) was developed. This review discusses, designing of the PMX, its application in clinical practice and the clinical outcomes.

Sepsis and septic shock remain important causes of morbidity and mortality in critical care practice [1]. In spite of aggressive treatment, sepsis produced multiorgan dysfunction. The Surviving Sepsis Campaign recommended the following international guidelines for management of severe sepsis and septic shock: (1) early goal-directed resuscitation of the septic patients during the first 6 h after recognition; (2) lung-protective ventilation; (3) after obtaining culture, administration of broad-spectrum antibiotics; (4) the use of activated protein C, and (5) other treatment including correction of anemia, the use of corticosteroid, and tight glucose control. [2]. In spite of these aggressive treatments, rates of sepsis-induced multiorgan dysfunction and mortality due to sepsis and septic shock still remain high. Recently, Cruz et al. [3] reported the results of the EUPHAS (Early Use of Polymyxin B Hemoperfusion in Abdominal Sepsis) study. This study tested whether polymyxin B convalently immobilized fiber (PMX) hemoperfusion improved mean arterial pressure (MAP) and less requirement for vasopressors in patients with septic shock due to presumed abdominal infections. In this preliminary study, PMX-filtration (PMX-F) improved hemodynamics and organ dysfunction and reduced 28-day mortality in these patients.

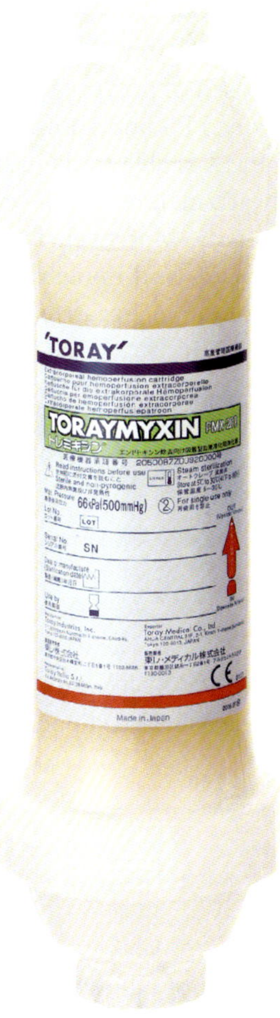

Fig. 1. The polymyxin B convalently immobilized fiber.

Before carrying out large multicenter studies, we would like to review (1) the design of the PMX, (2) its application in clinical practice, and (3) the clinical outcomes.

Designing PMX

The polymyxin column is shown in figure 1. Polymyxin B is an antibiotic agent that has a strong bactericidal activity to Gram-negative bacteria. The mechanisms of action are attributed to the electrostatic interaction of cationic antibiotic molecules with acidic phospholipids and the lipopolysaccharide (LPS) of the outer membrane of Gram-negative bacteria, which results in altered permeability of the outer membrane. PMX is immobilized convalently on the surface of a polystyrene-derived fibrous material, which is composed of polypropylene reinforced conjugated fibers. PMX absorbs many kinds of LPS and shows an endotoxin detoxification capacity. Besides this activity, PMX is able to neutralize endotoxin toxicity, including lethal toxicity, the Schwartzman reaction, certain blood coagulation defects and limulus gelation acitivity. Moreover, PMX remove cytokine inducing substances such as peptidoglycans and lipoteichoic acids, indicating that PMX might be applicable not only for Gram-negative sepsis, but also for Gram-positive sepsis [4].

The PMX cartridge is autoclave-sterilized, and the cartridge is filled with physiological saline. Application of a thin fibrous carrier allows carrying out a hemoadsorption cartridge with a large surface area and a low pressure drop in the blood-flow compartment.

Animal Experiments

In an endotoxin-challenged canine model, treatment with PMX showed quick recovery from severe hypotension and produced a longer survival compared to that without PMX [5].

Clinical Studies

Effects of PMX on Hemodynamics in Patients with Sepsis/Septic Shock

It is well known that endotoxin produces marked reduction of blood pressure by dilating the vascular beds. Using PMX cartridge, several clinical studies showed that blood pressure increased MAP on average by between 15 and 22 mm Hg. Application of PMX-F for patients with pre-PMX-F MAP of at least 70 mm Hg demonstrated a greater improvement in MAP. In addition to the improvement of MAP, a trend toward a decrease in the dose of vasoactive agents after treatment with PMX has been reported[6–10]. Suzuki et al. [11] compared the dose of dopamine between patients treated with continuous venovenous hemodiafiltration (CHDF) alone or PMX-F in combination. Patients with septic shock and treated with CHDF only needed a higher dose of dopamine at the end of the treatment. On the other hand, patients treated with PMX-F and CHDF in combination did not require the increased dose of dopamine.

PaO_2/FiO_2 Ratio

Previous studies [12–14] demonstrated that the PaO_2/FiO_2 ratio increased after treatment with PMX-F, probably due to improvement in MAP.

Endotoxin Levels
The effects of PMX on endotoxin levels have been reported to be reduced by between 33 and 80% from their pre-PMX levels. It remains uncertain whether the levels of endotoxin are reduced or not by treatment with PMX-F [7, 12].

Survival Rate
According to Cruz et al. [15], they reported that pooled mortality rates were 61.5% in the conventional therapy group and 33.5% in the PMX-F group. In the pooled estimate, PMX-F appeared to significantly reduce mortality compared with conventional medical therapy (RR 0.53, 95%CI 0.43–0.65). The results were similar between randomized controlled trials (RCT) and non-RCT and when 2 RCTs enrolling patients with MRSA infections were excluded (RR 0.55, 95%CI 0.40–0.78).

Nemoto et al. [16] reported that patients with a clinical diagnosis of sepsis, severe sepsis and septic shock according to the criteria of the American College of Chest Physicians/Society of Critical Care Medicine Consensus Conference [17] were randomly assigned to a group with or without PMX-F treatment. In their study, access to blood for direct hemoperfusion (DHP) with PMX-F adsorbent therapy was achieved via a double-lumen catheter inserted into the femoral vein of each patient by Seldinger's method. Immediately after ascertaining that the patients met the criteria of SIRS with infection, DHP with PMX-F was started. The time span between diagnosis and initiation of PMX-F treatment was usually less than 3 h. DHP was carried out for 4 h at a flow rate of 80–100 ml/min through a venovenous catheter similar to that used for acute dialysis. The anticoagulant was nafamostat mesilate (Torii Co. Ltd., Tokyo, Japan) and the usual doses were between 30 and 50 mg/h. DHP with PMX-F was performed once or twice, depending on the patient's condition. In their study, the overall survival rate was significantly improved in comparison with the control group (41 vs. 11%, $p = 0.002$). In patients with an Acute Physiology and Chronic Health Evaluation (APACHE) II score less than 20, treatment with PMX was shown to improve outcome (65 vs. 19%, $p = 0.01$). In cases of more severe sepsis with an APACHE II score of 20–29, PMX still maintained efficacy in improving outcome (40 vs. 11%, $p = 0.04$). However, PMX treatment did not improve the survival rate in patients with an APACHE II score of greater than 30 (survival rate 7 vs. 0%, $p = 0.59$; fig. 2). Pretreatment APACHE II scores were highly correlated with death among the patients given supportive therapy only in all populations examined. (with PMX-F, $r = 0.697$, $p < 0.0001$; without PMX-F, $r = 0.473$, $p < 0.001$; fig. 3).

Suzuki et al. [11] compared the effects of CHDF alone or CHDF and PMX in combination on survival rates of patients with sepsis and acute renal failure, the survival rate of all patients at 14 days was 25% for those with CHDF and 75% for those with PMX and CHDF treatment, and these were maintained at the end of the study (fig. 4).

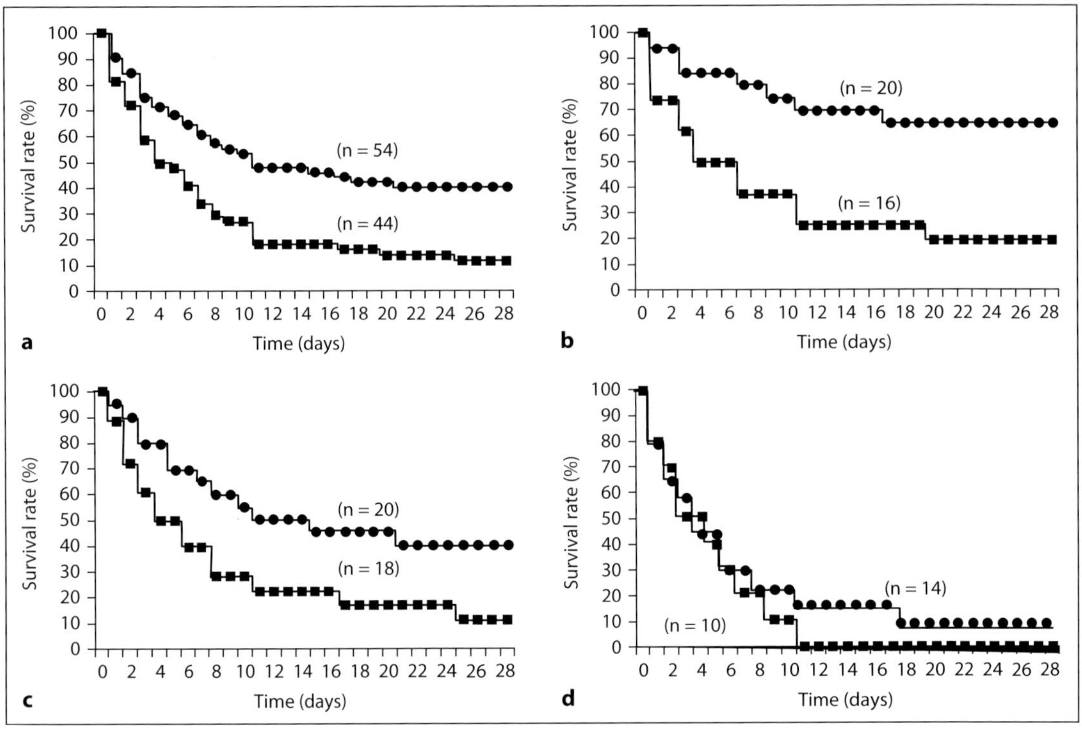

Fig. 2. Kaplan-Meier analysis of survival in patients with sepsis receiving PMX-F treatment or supportive care only. An intention-to-treat analysis of mortality from all casues at 28 days by the Cochran-Mantel-Haenazel test showed a significant reduction in mortality in patients treated with PMX-F. ■ = PMX-F–, ● = PMX-F+. **a** All patients with septic shock (n = 98). **b** Patients with mild sepsis (APACHE II score <20). **c** Patients with moderate sepsis (APACHE II score 20–29; n = 38). **d** Patients with severe sepsis (APACHE II score >30; n = 24).

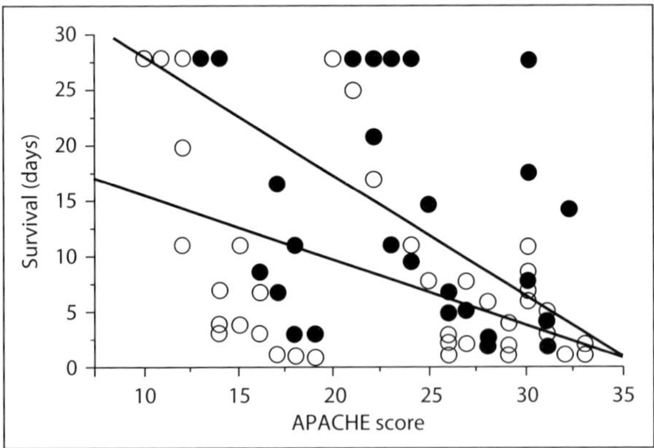

Fig. 3. Correlation between APACHE II scores (x axis) and survival time (y axis) in patients with ($y = 39.57 - 1.09x$; $r^2 = 0.486$) or without ($y = 21.56 - 0.58x$; $r^2 = 0.224$) PMX-F treatment. PMX-F treatment improved the survival rate. ○ = PMX-F–, ● = PMX-F+.

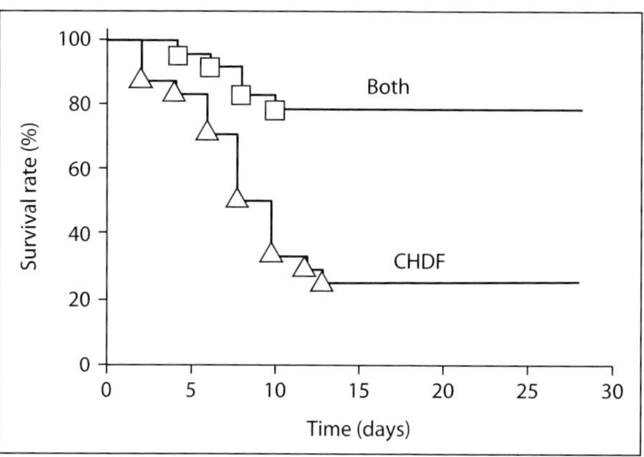

Fig. 4. Survival rate in patients with septic shock after treatment with PMX-F and CHDF in combination and CHDF alone during 28 days. Kaplan Meier survival curves. Combination treatment improved survival rate significantly compared to treatment with CHDF alone ($p < 0.01$). Comparison was made by the Cochran-Mantel-Haenszel test.

Drawbacks of Treatment with PMX-F

Treatment with PMX-F may have 2 major drawbacks in clinical practice. Firstly, it is necessary to draw blood for DHP with PMX-F adsorbent therapy. During this therapy, approximately 100 ml of blood circulates extracorporally. To apply this technique, a double-lumen catheter needs to be inserted into the femoral vein by Seldinger's method. Secondly, anticoagulant therapy is needed to apply this treatment. Compared to therapy with the infusion of agents that bind to and neutralize endotoxin, treatment with PMX-F needs manpower, machines and catheters. This may result in an economic constraint to the application of this therapy. Up to now, however, there have been few successful methods for treating patients with sepsis or septic shock. Therefore, for the time being, this method will be helpful and valuable for the improvement of survival of patients with severe sepsis.

RCT of Treatment with PMX-F in Patients with Sepsis

Cruz et al. [3] carried out a prospective, multicenter, RCT (Early Use of Polymixn B Hemoperfusion in Abdominal Sepsis, EUPHAS). Sixty-four patients were enrolled with severe sepsis or septic shock who underwent emergency surgery for intra-abdominal infection. Patients were randomized to either conventional therapy or conventional therapy plus 2 sessions of PMX-F. In their results, MAP increased (76–84 mm Hg, $p = 0.001$) and vasopressor requirement decreased (inotropic score 29.9–6.8, $p < 0.001$) at 72 h in the PMX-F group but not in the conventional therapy group. The Pao_2:Fio_2 ratio increased slightly (235–264, p = 0.79). SOFA scores improved in the PMX-F group but not in the conventional therapy group (change in SOFA, –3.4 vs. –0.1, p < 0.01), and 28-day mortality was 32% (11/34 patients) in the PMX-F group and 53% (16/30 patients) in the conventional therapy group (adjusted HR 0.36, 95%CI 0.16–0.80). Their findings were similar, in that the PMX-F group had an adjusted HR of 0.36 for 28-day mortality and 0.43 for all-cause hospital mortality. The delta SOFA scores were significantly better in the PMX-F group, indicating improvement in overall organ function, particularly in the cardiovascular component. Improvement in blood pressure and reduction in vasopressor doses have also been demonstrated in other studies. In our study, even as the dose of vasoactive agents (indicated by the inotropic score) was reduced, there was a significant increase in MAP in the PMX-F group at 72 h. Accordingly, the vasopressor dependency index decreased significantly in the PMX-F group but not in the conventional group.

In terms of pulmonary function, there was no significant difference in mechanical ventilation-free days or delta respiratory SOFA score between the 2 groups. Observational studies and case reports have suggested a beneficial effect of PMX-F therapy on Pao_2/Fio_2 ratio in sepsis, acute lung injury and acute respiratory diseases syndrome.

In term of renal dysfunction, the delta renal SOFA score at 72 h was better in the PMX-F group, indicating some improvement in the degree of renal organ dysfunction in this group. However, the proportion of patients treated with RRT was similar between the 2 groups. Earlier studies have demonstrated positive renal effects of PMX therapy.

Conclusion

Direct hemoperfusion with PMX-F has favorable effects on blood pressure, vasoactive pressor agents, Pao_2/Fio_2 ratio. These produce improvement in hemodynamics and organ dysfunction and reduce mortality in patients suffering from severe sepsis.

References

1 Joannidis M: Continuous renal replacement therapy in sepsis and multisystem organ failure. Semin Dial 2009;22:160–164.

2 Dellinger RP, Levy MM, Carlet JM, et al: Surviving Sepsis Campaign: international guidelines for management of severe sepsis and septic shock: 2008. Intensive Care Med 2008;34:17–60.

3 Cruz DN, Antonelli M, Fumagalli R, et al: Early use of polymyxin B hemoperfusion in abdominal septic shock: the EUPHAS randomized controlled trial. JAMA 2009;301:2445–2452.

4 Shoji H, Tani T, Hanasawa K, Kodama M: Extracorporeal endotoxin removal by polymyxin B immobilized fiber cartridge: designing and antiendotoxin efficacy in the clinical application. Ther Apher 1998;2:3–12.

5 Aoki H, Kodama M, Tani T, Hanasawa K: Treatment of sepsis by extracorporeal elimination of endotoxin using polymyxin B-immobilized fiber. Am J Surg 1994;167:412–417.
6 Nakamura T, Kawagoe Y, Matsuda T, et al: Effect of polymyxin B-immobilized fiber on blood metalloproteinase-9 and tissue inhibitor of metalloproteinase-1 levels in acute respiratory distress syndrome patients. Blood Purif 2004;22:256–260.
7 Tani T, Hanasawa K, Endo Y, et al: Therapeutic apheresis for septic patients with organ dysfunction: hemoperfusion using a polymyxin B immobilized column. Artif Organs 1998;22:1038–1044.
8 Ueno T, Sugino M, Nemoto H, Shoji H, Kakita A, Watanabe M: Effect over time of endotoxin adsorption therapy in sepsis. Ther Apher Dial 2005;9:128–136.
9 Ono S, Tsujimoto H, Matsumoto A, Ikuta S, Kinoshita M, Mochizuki H: Modulation of human leukocyte antigen-DR on monocytes and CD16 on granulocytes in patients with septic shock using hemoperfusion with polymyxin B-immobilized fiber. Am J Surg 2004;188:150–156.
10 Vincent JL, Laterre PF, Cohen J, et al: A pilot-controlled study of a polymyxin B-immobilized hemoperfusion cartridge in patients with severe sepsis secondary to intra-abdominal infection. Shock 2005;23:400–405.
11 Suzuki H, Nemoto H, Nakamoto H, et al: Continuous hemodiafiltration with polymyxin-B immobilized fiber is effective in patients with sepsis syndrome and acute renal failure. Ther Apher 2002;6:234–240.
12 Tojimbara T, Sato S, Nakajima I, Fuchinoue S, Akiba T, Teraoka S: Polymyxin B-immobilized fiber hemoperfusion after emergency surgery in patients with chronic renal failure. Ther Apher Dial 2004;8:286–292.
13 Tsushima K, Kubo K, Koizumi T, et al: Direct hemoperfusion using a polymyxin B immobilized column improves acute respiratory distress syndrome. J Clin Apher 2002;17:97–102.
14 Kushi H, Nakahara J, Miki T, Okamoto K, Saito T, Tanjo K: Hemoperfusion with an immobilized polymyxin B fiber column inhibits activation of vascular endothelial cells. Ther Apher Dial 2005;9:303–307.
15 Cruz DN, Perazella MA, Bellomo R, et al: Effectiveness of polymyxin B-immobilized fiber column in sepsis: a systematic review. Crit Care 2007;11:R47.
16 Nemoto H, Nakamoto H, Okada H, et al: Newly developed immobilized polymyxin B fibers improve the survival of patients with sepsis. Blood Purif 2001;19:361–368; discussion 368–369.
17 Bone RC, Balk RA, Cerra FB, et al: Definitions for sepsis and organ failure and guidelines for the use of innovative therapies in sepsis. The ACCP/SCCM Consensus Conference Committee. American College of Chest Physicians/Society of Critical Care Medicine. Chest 1992;101:1644–1655.

Hiromichi Suzuki, PhD, MD
Department of Nephrology, Saitama Medical University
Morohongo 38, Moroyamamachi Irumagun
Saitama 350-0495 (Japan)
Tel. +81 492 76 1620, Fax +81 492 76 1620, E-Mail iromichi@saitama-med.ac.jp

Current Progresses in Methodology in Blood Purification in Critical Care

Suzuki H, Hirasawa H (eds): Acute Blood Purification.
Contrib Nephrol. Basel, Karger, 2010, vol 166, pp 158–166

Special Considerations in Continuous Hemodiafiltration with Critically Ill Pediatric Patients

Hidetoshi Shiga[a] · Yoshihiko Kikuchi[a] · Noriyuki Hattori[b] · Hiroyuki Hirasawa[b]

[a]Emergency and Intensive Care Center, Teikyo University Chiba Medical Center, Ichihara, and [b]Deparment of Emergency and Critical Care Medicine, Chiba University Graduate School of Medicine, Chiba, Japan

Abstract

Continuous hemodiafiltration (CHDF) has become an essential procedure in critical care. However, effective application of this modality to pediatric patients is associated with several problems derived from their smaller body size and weight compared with adults. We have successfully conducted CHDF in pediatric patients, even newborns, by taking such problems into consideration and navigating around them. Successful CHDF in pediatric patients was achieved by careful and exact execution of the following countermeasures to overcome pediatric-specific problems: minimization of the priming volume; use of colloid solutions or whole blood as priming solution; maintaining secure vascular access; selection of an appropriate anticoagulant; temperature control of both the patient's body and components of the hemofiltration circuit. In pediatric critical care, CHDF is safe and expected to demonstrate clinical efficacy across a wide spectrum of clinical problems, just as in adults.

Continuous hemodiafiltration (CHDF) is a safe therapeutic modality for critically ill adult patients with unstable hemodynamics and is expected to demonstrate clinical efficacy across a wide spectrum of clinical problems. It has become to be an essential procedure in critical care today [1–3]. However, application of this hemopurification therapy to pediatric patients is associated with several problems originating from their smaller body size and weight. We have successfully conducted CHDF in pediatric patients, including newborns, by taking such problems into consideration and developing strategies to deal with them [4]. In this way, we have safely and effectively introduced CHDF in pediatric critical care into our hospital.

Advantages of CHDF over Intermittent Hemodialysis

The advantages of CHDF over intermittent hemodialysis are as follows.

- it can be performed with a simple bedside console, without a complicated dialysate supply system;
- the influence of CHDF on the hemodynamics is small;
- it has excellent efficiency in the elimination of waste substances which are spreading widely throughout the body;
- it can help maintain homeostasis;
- the rate of adjustment of the blood condition is mild.

CHDF Conditions for Pediatric Patients

We use the JUN-505 (Junken Medical, Tokyo, Japan) as a bedside console, which has been specially developed for CHDF. Standard blood flow in pediatric patients, filtrate flow and dialysate flow are 1–1.5 ml/kg/min, 3–20 and 10–20 ml/kg/h, respectively.

Efficacy and Indications for CHDF in Pediatric Patients

The clinical uses of CHDF in pediatric patients with renal failure are almost the same as in adults [1, 2, 5]: (1) strict control of water, electrolyte and acid-base balance; (2) removal of metabolic waste products; (3) good nutrition management through removal of excess water given as carrier of total parenteral nutrition solution; (4) treatment of organ failure and prevention of its exacerbation through removal of humoral mediators, including cytokines; (5) improvement in respiratory functions by eliminating pulmonary interstitial edema; (6) removal of hepatic toxins, inducing hepatic coma [6]; (7) prevention of adverse effects upon plasma exchange.

This wide spectrum of clinical efficacy of CHDF in pediatric patients suggests the following indications for this therapy: acute renal failure, acute hepatic failure, acute respiratory failure, congestive heart failure, multiple organ failure, sepsis and rapidly exacerbating inborn errors of metabolism (such as lactic acidosis and hyperammonemia [7], rhabdomyolysis and HUS/TTP [8]).

Problems of CHDF in Pediatric Patients

We have demonstrated the clinical efficacy and safety of CHDF in critical care [1, 2, 5]. However, application of this hemopurification therapy to pediatric patients is associated with several problems derived from their smaller body size compared with adults.

Problems associated with CHDF in pediatric patients include the following:

- the volume of circulating plasma is smaller compared with adults, which results in a larger proportion of the priming volume of the CHDF circuit to the volume of circulating plasma, potentially generating a greater influence on the patient's circulation when CHDF is started or blood is returned to the body at the end of therapy;
- even a slight water imbalance may cause greater instability in hemodynamics compared with adults;
- a smaller blood vessel diameter makes establishment of vascular access more difficult [9];
- application of high blood flow rates in the extracorporeal circuit is difficult in pediatric patients due to their smaller body size and weight compared with adults;
- prevention of the above-mentioned blood coagulation within the CHDF circuit needs a relatively large dose of an anticoagulant;
- pediatric patients are more susceptible to hypothermia due to their smaller body size and weight compared with adults, and due to lesser basal heat generation resulting from fixed radiant heat loss.

Due to concern about these problems related to CHDF in pediatric patients, medical institutions not familiar with hemopurification therapies in such patients have conventionally preferred peritoneal dialysis [10–12], which utilizes the patient's peritoneum as a dialysis membrane and involves no extracorporeal circulation. This therapy, however, has several disadvantages: (1) insufficiency and uncertainty in water and solute removal; (2) inability to accurately control the volume of water removed; (3) a potential risk of intra-abdominal infections. Accordingly, reports of aggressive CHDF in pediatric patients have recently been accumulated [3, 7, 8, 13].

Countermeasures to Problems Associated with CHDF in Pediatric Patients

The following are our countermeasures to individual problems associated with CHDF in pediatric patients.

Priming Volume

A smaller body weight and volume of circulating plasma in pediatric patients compared with adults results in a larger relative proportion of priming volume (the volume within the hemofilter and the blood circuit) to the volume of circulating plasma in the body. This causes a drastic change in volume of circulating plasma upon both entry of the blood to the circuit at the start of CHDF and the blood return at the end of CHDF, inducing great fluctuations in hemodynamics.

To overcome this problem, we devised a extracorporeal hemofiltration circuit specially designed for pediatric use, involving a hemofilter with a smaller membrane area (0.1–0.3 m^2) and smaller priming volume, a blood circuit constructed with tubes of a smaller diameter and shorter length, and air trap chambers with a minimum volume.

Use of this circuit achieved a priming volume of 50–70 ml, and thereby minimized fluctuations in hemodynamics.

A crystalloid solution with no oncotic pressure, such as saline, used for priming of the hemofilter and the blood circuit induces a fall in oncotic pressure of plasma due to dilution immediately after the start of CHDF, which causes water shift to the extravascular space leading to edema formation and resultant hypovolemia. In critically ill patients with extremely unstable hemodynamics, such hypovolemia should be prevented by priming the hemofiltration circuit with colloid solutions (an albumin preparation or fresh frozen plasma) or whole blood. When using priming solutions such as fresh frozen plasma or whole blood, concomitant introduction of a large amount of electrolytes (such as sodium and potassium) and residual acid-citrate dextrose solution being added as anticoagulant may sometimes cause adverse effects. To solve this problem, we formed a shortcut within the extracorporeal hemofiltration circuit and allowed the priming solution to circulate within this shortcut circuit under a continuous supply of the dialysate for a certain period before the entire hemofiltration circuit was actually connected to the patient. Under the operation conditions (hemofilter: polymethylmethacrylate 0.3m^2, priming with 75 ml whole blood, blood flow = 30 ml/min, dialysis fluid flow = 500 ml/h, filtration rate = 0 ml/h, anticoagulant: nafamostat mesilate 3 mg), a baseline potassium level exceeding 8 mEq/l was reduced to within the normal range when we dialyzed the priming solution using the shortcut circuit for 5 min [4]. A 15-min predialysis may be sufficient to correct other parameters as well.

Transient hypervolemia occurring when blood within the hemofiltration circuit was returned to the patient upon circuit exchange was avoided by conducting both blood withdrawal to fill a new circuit and blood return from the one completing the hemofiltration session simultaneously to minimize fluctuations in both circulatory blood volume and extracorporeal volume.

When the patient could stand without CHDF until the new hemofiltration circuit was ready, blood within the former hemofiltration circuit was collected in a reservoir bag. The new hemofiltration circuit was primed with the collected blood. We formed a shortcut using the reservoir bag within the extracorporeal hemofiltration circuit and allowed the priming blood to circulate within this

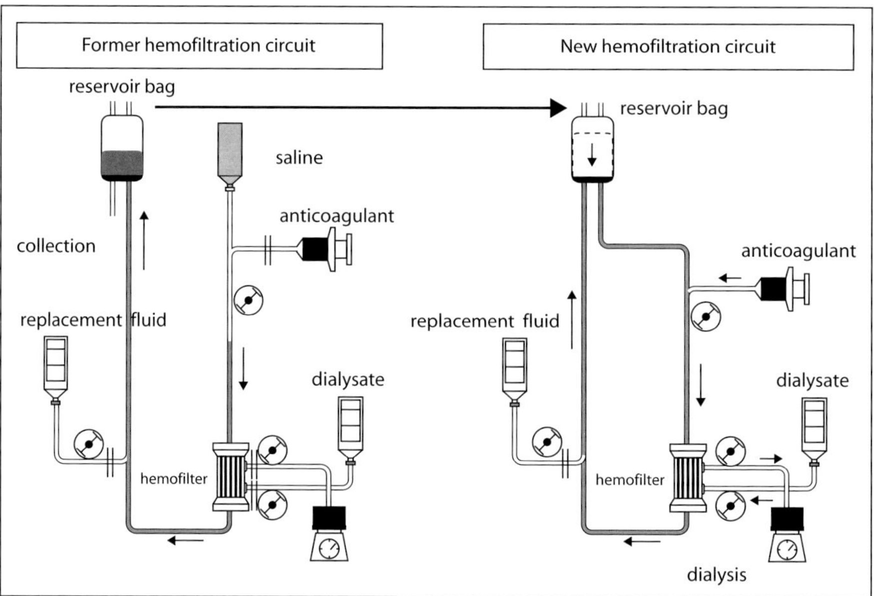

Fig. 1. Extracorporeal hemofiltration circuits' exchange without homologous blood transfusion.

shortcut circuit under a continuous supply of the dialysate for a certain period before the entire hemofiltration circuit was actually connected to the patient. The reservoir bag allowed the priming blood, which was diluted by rinsing saline, to be concentrated with filtration. With this method, we could exchange the extracorporeal hemofiltration circuits without homologous blood transfusion (fig. 1)

Influence of Water Imbalance on Hemodynamics

Since the volume of circulating plasma in pediatric patients is smaller than with adults, even a slight water imbalance may cause greater instability in hemodynamics. As a consequence, successful CHDF in pediatric patients requires stricter control of water balance at shorter intervals compared with adult cases. Also, any water removal/supply should be carefully performed at a low flow rate, taking enough time. Except for some special cases, a specially designed bedside console is used for control of blood flow rate, dialysate flow rate and filtration rate. This device allows accurate real-time control of water balance.

Vascular Access

Securing vascular access to ensure stable blood flow is the first prerequisite of successful hemopurification therapy involving extracorporeal circulation including CHDF, especially on pediatric patients.

Table 1. Vascular access catheters in pediatric patients aged <1 year

Patient No.	Time after birth, days	Sex	Body weight, kg	Disease	Catheter size, F	Insertion site
1	8	M	2.67	CDH	7	right IJV
2	5	F	1.60	CDH	7	right IJV
3	289	M	9.40	VSD, ARF	9	right IJV
4	231	M	6.80	liver transplantation	9	right IJV
5	1	F	1.46	lactic acidosis	9	right IJV
6	4	M	4.10	septic shock	7	right IJV
7	15	M	2.80	amniocele	7	left FV
8	33	F	3.11	duodenal stenosis	7	right IJV
9	15	M	2.11	lactic acidosis	6	right IJV

Further data (means ± SD): blood flow 8.77 ± 1.33 ml/kg/min; filter life 26.3 ± 9.20 h; CHDF duration 9.83 ± 10.4 days (2–31 days). CDH = Congenital diaphragmatic hernia; IJV = internal jugular vein; VSD = ventricular septal defect; ARF = acute renal failure; FV = femoral vein.

In general, CHDF in older children weighing ≥40 kg was performed in a venovenous mode, with vascular access provided via a 12-french vascular access catheter (the same size as used in adult cases) placed in the internal jugular vein or the femoral vein applying the Seldinger technique.

The procedure applied to patients weighing <40 kg was the same as that described above, except that a 6- to 9-french vascular access catheter was used in place of a 12-french catheter. We preferred CHDF in a venovenous mode whenever a sufficient blood flow rate was achieved via a flexible double-lumen catheter placed in the internal jugular vein, the femoral vein or the umbilical vein, regardless of the patient's body weight and even in newborns. The youngest patient with the smallest body weight in whom this procedure has successfully been conducted is a newborn baby weighing 1.46 kg treated within 1 day of birth. Table 1 shows vascular access catheters in patients less <1 year old.

A catheter with a maximum diameter and a minimum total length that ensure maximum blood flow rate with minimum fluid resistance is desirable for providing vascular access. Catheters currently commercially available for this purpose are divided into 2 groups, each involving a different type of manipulation for insertion into the blood vessel. Catheters of one type are inserted by passing through an outer cylinder introduced into the desired blood vessel by puncture. Catheters of the other type, designed for the Seldinger technique, are inserted by introducing a guide wire first and then the catheter itself. Use of catheters with a guide wire allows insertion of a catheter with a larger diameter, even when the needle that was first punctured is the same, and is therefore more preferable in CHDF in pediatric patients. In some cases ultrasonography-guided puncture of vein is preferable.

Blood Coagulation within the Hemofiltration Circuit

The blood flow rate during a CHDF session was maintained at approximately 1 ml/kg/min or higher, considering the unstable hemodynamics. Since the risk of blood coagulation within the hemofiltration circuit increases at lower blood flow rates, special care was taken to obtain a blood flow rate exceeding a minimum acceptable value of 20 ml/min. Doses of anticoagulants were strictly controlled to minimum essential levels required for sufficient anticoagulation.

Anticoagulants

Administration of anticoagulants in therapies involving extracorporeal circulation is essential for the prevention of blood coagulation within the blood circuit, but this increases a tendency towards bleeding that results in bleeding from the vascular access sites, the gastrointestinal tracts and the surgical wounds. Nafamostat mesilate, a synthetic protease inhibitor with anticoagulation properties and the standard anticoagulant in CHDF in adults [14, 15], was also used as the first-line anticoagulant in pediatric patients. The dose of this anticoagulant was adjusted to maintain an activated coagulation time of approximately 150 s measured for blood in systemic circulation during a CHDF session, since we have reported that this period of time produces the most effective anticoagulation with the minimum risk of bleeding complications in adult patients [14, 15]. Although the dose of nafamostat mesilate depended on factors such as the patient's conditions and blood flow rate, 0.1–1.0 mg/kg/h was usually sufficient to perform CHDF safely without serious hemorrhagic complications.

When a high dose of nafamostat mesilate failed to eliminate frequent blood clotting within the hemofiltration circuit, low-molecular-weight heparin [16] was used in combination. Low-molecular-weight heparin is associated with a greater risk of hemorrhagic complications compared with nafamostat mesilate, but has an economical advantage.

Body Temperature

Pediatric patients may be susceptible to hypothermia caused by a large volume of extracorporeal circulation, due to their smaller body size and weight compared with adults [17]. Every possible care should be taken to eliminate any cause of heat loss from the patient's body during CHDF, including the following: (1) warming both dialysate and replacement fluids; (2) wrapping the tube line and hemofilter with aluminum foil to keep it warm (the foil can hold air surrounding the tube line and hemofilter inside the foil, therefore preventing radiant heat loss and convective heat loss); we had patients who did not develop hypothermia with this aluminum foil, but did develop hypothermia without it; (3) putting a coiled tube in the hemofiltration blood circuit and submerging it in a heated bath; (4) using a bed warmer to warm the patient's body.

With these procedures, CHDF was safely performed without serious complications in pediatric patients and even in newborns.

In summary, CHDF can safely be used for critically ill adult patients with unstable hemodynamics and is expected to demonstrate clinical efficacy across wide spectrum of disease. Here, CHDF was also safely conducted in pediatric patients, including newborns, without serious complications and showing a similar spectrum of clinical efficacy. The essential prerequisite for successful CHDF in pediatric patients was to consider some special problems originating from the smaller body size and weight of pediatric patients compared with adults.

Conclusions

Successful CHDF in pediatric patients was achieved by careful and exact execution of the following countermeasures to overcome problems specific to pediatric patients: minimization of the priming volume; use of colloid solutions or whole blood as priming solution; maintaining secure vascular access; selection of an appropriate anticoagulant; temperature control of both the patient's body and components of the hemofiltration circuit.

References

1 Nakada T, Oda S, Matsuda K, Sadahiro T, Nakamura M, Abe R, Hirasawa H: Continuous hemodiafiltration with PMMA hemofilter in the treatment of patients with septic shock. Mol Med 2008;14:257–263.

2 Oda S, Hirasawa H, Shiga H, Nakanishi K, Matsuda K, Nakamura M: Continuous hemofiltration/hemodiafiltration in critical care. Ther Apher 2002;6:193–198.

3 Dunham CM: Clinical impact of continuous renal replacement therapy on multiple organ failure. World J Surg 2001;25:669–676.

4 Shiga H, Hirasawa H, Oda S, Matsuda K,Ueno H, Nakamura M: Continuous hemodiafiltration in pediatric critical care patients. Ther Apher Dial 2004;8:390–397.

5 Matsuda K, Hirasawa H, Oda S, Shiga H, Nakanishi K: Current topics on cytokine removal technologies. Ther Apher 2001;5:306–314.

6 Yokoi T, Oda S, Shiga H, Matsuda K, Sadahiro T, Nakamura M, Hirasawa H: Efficacy of high-flow dialysate continuous hemodiafiltration in the treatment of fulminant hepatic failure. Transfus Apher Sci 2009;40:61–70.

7 Schaefer F, Straube E, Oh J, Mehls O, Mayatepek E: Dialysis in neonates with inborn errors of metabolism. Nephrol Dial Transplant 1999;14:910–918.

8 Ponikvar P, Kandus A, Urbanci A, Kornhauser AG, Primozic J, Ponikvar JB: Continuous renal replacement therapy and plasma exchange in newborns and infants. Artif Organs 2002;26:163–168.
9 Chand DH, Valentini RP, Kamil ES: Hemodialysis vascular access options in pediatrics: considerations for patients and practitioners. Pediatr Nephrol 2009;24: 1121–1128.
10 Golej J, Kitzmueller E, Hermon M, Boigner H, Burda G, Trittenwein G: Low-volume peritoneal dialysis in 116 neonatal and paediatric critical care patients. Eur J Pediatr 2002; 161:385–389.
11 Leonard MB, Donaldson LA, Ho M, Geary DF: A prospective cohort study of incident maintenance dialysis in children: an NAPRTC study. Kidney Int 2003;63:744–755.
12 Strazdins V, Watson AR, Harvey B: Renal replacement therapy for acute renal failure in children: European guidelines. Pediatr Nephrol 2004;19:199–207.
13 Walters S, Porter C, Brophy PD: Dialysis and pediatric acute kidney injury: choice of renal support modality. Pediatr Nephrol 2009; 24:37–48.
14 Matsuo T, Kario K, Nakao K, Yamada T, Matsuo M: Anticoagulation with nafamostat mesylate, a synthetic protease inhibitor, in hemodialysis patients with a bleeding risk. Haemostasis 1993;23:135–141.
15 Amanzadeh J, Reilly RF Jr: Anticoagulation and continuous renal replacement therapy. Semin Dial 2006;19:311–316.
16 Spiegel DM, Anderson RJ: Is low-molecular-weight heparin useful for venovenous hemofiltration in the intensive care unit? Crit Care Med 1999;27:2316–2317.
17 Kornecki A, Tauman R, Lubetzky R, Sivan Y: Continuous renal replacement therapy for non-renal indications: experience in children. Isr Med Assoc J 2002;4:345–348.

Hidetoshi Shiga
Emergency and Intensive Care Center
Teikyo University Chiba Medical Center
3426-3 Anesaki, Ichihara, Chiba 299-0111, Japan
Tel. +81 436 621211, Fax +81 436 621327, E-mail shiga@med.teikyo-u.ac.jp

Suzuki H, Hirasawa H (eds): Acute Blood Purification.
Contrib Nephrol. Basel, Karger, 2010, vol 166, pp 167–172

Selection of Modality in Continuous Renal Replacement Therapy

Yoshihiko Kanno · Hiromichi Suzuki

Department of Nephrology, School of Medicine, Faculty of Medicine, Saitama Medical University, Saitama, Japan

Abstract

Continuous hemoperfusion therapies are now widely used in critical care, and could prove to be life-saving for patients unable to receive regular hemoperfusion treatments. Unfortunately, due to the inherent difficulties in assessing the effects of treatment upon critically ill patients, the efficacy of this modality has yet to be proven.

Instead of focusing exclusively on a particular form of continuous hemoperfusion or a direct comparison between the different types available, this report provides a general overview of the studies reporting on its efficacy across a wide range of conditions. The authors conclude that continuous hemoperfusion could be beneficial in some cases, but this is highly dependent upon the particular modality used.

Modality in Continuous Renal Replacement Therapy (CRRT)

Continuous procedures of hemoperfusion have several modalities, and common potential advantages when compared with intermittent hemoperfusion procedures. In fact, they may change the long-term prognosis in critically ill patients [1]. Continuous procedures are hemodynamically stable because of the lower rate of fluid removal and control of azotemia. Sometimes, the nomenclatures for these procedures are confusing as many variations are carried out. For the purposes of this chapter, we will use the terminology set out below.

- Continuous hemofiltration (CHF). A large volume of replacement fluid is infused into either the inflow or the outflow blood line instead of dialysis solution. The reason for a large volume is that the volume of fluid that needs to be ultrafiltered across the membrane includes both replacement and excess fluid.

- Continuous hemodialysis (CHD). Dialysis solution is passed through the dialysate compartment of the filter continuously and at a slow rate. Diffusion is the primary method of solute removal. The amount of fluid that must be ultrafiltered across the membrane is low (3–6 l/day) and limited to excess fluid removal.
- Continuous hemodiafiltration (CHDF). This method combines CHD and CHF. Thus, dialysis solution is used, and replacement fluid is also infused into either the inflow or the outflow blood line. The daily volume of fluid that is ultrafiltered across the membrane is higher than with CHD, but not as high as with CHF.
- Slow continuous ultrafiltartion (SCUF). Also called continuous extracorporeal ultrafiltration method (C-ECUM), it simply removes the excessive volume by ultrafiltration without dialysis solution and replacement fluid.

Selection of Modality in CRRT: Survival in ICU Acute Renal Failure

Acute kidney injury in ICU is the most frequent status that requires CRRT. Therefore, many clinical trials have reported and compared the effect of CRRT in several modalities. As the conditions of each study – including indication, protocol, modality of CRRT and patient status – are different, it is difficult to compare them directly. However, many trials suggested that the modality of dialysis therapy did not effect the prognosis of the patient with acute kidney injury in ICU. Bagshaw et al. [2] showed in their meta-analysis no statistical evidence that initial modality influenced mortality (OR 0.99, nine trials, n = 1,403) or recovery to renal replacement therapy independence (OR 0.76, four trials, n = 306). Augustine et al. [3] carried out a randomized control trial that compared CHD with intermittent hemodialysis (IHD). There were no differences in survival or renal recovery between groups, though there was greater net volume removal in the CVVHD group during the first 72 h [3]. Data from the Program to Improve Care in Acute Renal Disease (PICARD), a multicenter observational study of acute kidney injury, were analyzed by Cho et al. [4] with surprising results. Among 398 patients who required dialysis, the risk for death within 60 days was examined by assigned initial dialysis modality [CRRT (n = 206) vs. IHD (n = 192)]. Crude survival rates were lower for patients who were treated with CRRT than IHD (survival at 30 days: 45 vs. 58%, p = 0.006). Adjusted for a several clinical factors, the relative risk for death associated with CRRT was also 1.82. Therefore, the authors had to conclude that CRRT was associated with increased mortality among critically ill patients with acute kidney injury. As shown in table 1, other studies [5–7] suggested that continuous therapy did not provide better prognosis, except 1 old study that reported CHDF reduces morbidity and mortality compared with conventional dialysis therapy [8]. Not only modality of CRRT, but also effluent

Table 1. Several studies comparing continuous and intermittent therapy for acute renal failure

Study	Design	Therapy	Effect of modality
Lins, 2009 [7]	RCT	CHD (n = 172) and IHD (n = 144)	No effect on mortality
Palevsky, 2008 [19]		CHD and IHD (n = 1,124)	No effect on mortality
Rabindranath, 2007 [20]	Meta-analysis	CHD and IHD (n = 1,550)	No effect on mortality
Uehlinger, 2005 [21]		CHDF (n = 70) and IHD (n = 55)	No effect on mortality
Uchino, 2007 [22]	Cohort	CHDF (n = 1,006) and IHD (212)	No effect on mortality and hospital discharge; CHDF increased renal recovery rate.
Vinsonneau, 2006 [23]	RCT	CHDF and IHD (n = 350)	No effect on mortality

Only the name of the first author for each study is shown.

flow rate did not effect on the mortality of the patients [9]. Bellomo et al. [8] compared the effluent flow of either 40 ml/kg body weight/h (higher intensity) or 25 ml with 747 and 761 patients. At 90 days after randomization, 322 deaths had occurred in the 40 ml group and 332 deaths in the 25 ml group, for a mortality of 44.7% in each group (OR 1.00, p = 0.99). Tolwani et al. [10] also concluded that there is no significant difference in survival of acute renal failure patients receiving CHDF with prefilter replacement fluid at an effluent rate of either 35 ml/kg/h or 20 ml/kg/h. Other notable trials are listed in table 1.

Efficacy of CRRT in Other Conditions

Burns

Leblanc et al. [11] investigated 16 burned patients with acute renal failure who received CRRT. The patients received CHD and CHDF and were compared with 33 patients without burns who were treated for acute renal failure. Although bleeding complications were more frequent in burned patients (56 vs. 15%), mortality rates were similar in both groups (82% for both).

Sepsis

Our retrospective study compared the efficacy of polymyxin-B immobilized fiber (PMX-F) alone and in combination with CHF on the prognosis of 48 critically ill patients with sepsis [12]. We divided these patients into 2 groups according to their primary diseases (cardiovascular, gastrointestinal). The survival rate differed significantly between the 2 groups. The patients in the cardiovascular group survived longer than those in the gastrointestinal group. Moreover, for the patients with cardiovascular disease, there was no significant difference in the survival rate between treatment with PMX-F alone and with PMX-F and CHF in combination. In contrast, for the patients with gastrointestinal disease, there was a significant difference between treatment with PMX-F alone and with PMX-F and CHF in combination. Primary disease would be one useful criterion to whether PMX-F should be given alone or in combination with CHF.

Cardiac Status

As shown above, modality of dialysis seems not to improve the prognosis of patients with acute renal failure. However, as several studies suggest that continuous therapy could provide some advantages for cardiac status, John et al. [13] compared the effects of CHF and IHD in septic shock patients prospectively. In patients who received CHF ($n = 20$), systolic blood pressure increased significantly 2 h after the initiation of therapy, compared with patients who received IHD ($n = 10$), although there was no statistically significant difference in mortality between them. Davenport et al. [14] carried out another prospective randomized controlled trial to investigate the effects on cardiovascular stability. In 32 patients with acute renal failure receiving intermittent hemofiltration (IHF) or CHDF, cardiac output, tissue oxygen delivery, and uptake were compared. During the first hour of treatment, there was a reduction in cardiac index of 15 ± 2% during IHF compared with no significant change during CHDF (3 ± 3%; $p < 0.05$). This reduction in cardiac output during IHF was associated with a maximum reduction in mean arterial pressure from 82 ± 2 to 66 ± 2 mm Hg ($p < 0.001$). The authors stated that the use of continuous forms of renal replacement therapy is preferred for its improved cardiovascular tolerance compared with daily intermittent treatments.

Volume Control

Baldwin et al. [15] compared IHF and CHF from the standpoint of fluid control. In 16 critically ill patients with acute renal failure, adequate prescribed fluid removal was achieved with both techniques. However, as expected, fluid was removed at a faster rate during IHF. This was initially associated with a lower blood pressure than during CHF, where blood pressure increased.

Miscellaneous Trials

Control of azotemia was investigated by Bellomo et al. [16], who looked at 47 critically ill patients with multiorgan failure and acute renal failure treated with IHD and 47 similar patients treated with CHDF. They found that CHDF was associated with significantly lower plasma urea ($p < 0.0001$) and serum creatinine ($p < 0.01$) levels at 24 h of treatment despite similar levels at the start of therapy. However, Clark et al. [17] did not see any difference between CHF and IHD with their same scale study. For acid-base balance, Barenbrock et al. [18] recommended a biocarbonate buffered replacement to normalize acidosis of patients without the risk of alkalosis. Their data also suggested that the use of bicarbonate during CHF reduces cardiovascular events in critically ill patients with acute renal failure, compared with lactate buffered replacement.

Conclusions

Continuous therapy is now widely used in critical scenes, and can rescue patients who cannot receive regular hemoperfusion treatment. It is very difficult to compare treatment for patients with critical status, and – regrettably – the advantage of this useful therapy has not been proved in any modality for the prognosis of patients. As shown in this chapter, it is likely that continuous therapy could be of benefit in at least some conditions. Thus, adequate selection of modality in continuous therapy is necessary to gain the better results in patient prognosis.

References

1 Triverio PA, Martin PY, Romand J, Pugin J, et al: Long-term prognosis after acute kidney injury requiring renal replacement therapy. Nephrol Dial Transplant 2009;24:2186–2189.

2 Bagshaw SM, Berthiaume LR, Delaney A, Bellomo R: Continuous versus intermittent renal replacement therapy for critically ill patients with acute kidney injury: a meta-analysis. Crit Care Med 2008;36:610–617.

3 Augustine JJ, Sandy D, Seifert TH, Paganini EP: A randomized controlled trial comparing intermittent with continuous dialysis in patients with ARF. Am J Kidney Dis 2004;44:1000–1007.

4 Cho KC, Himmelfarb J, Paganini E, Ikizler TA, et al: Survival by dialysis modality in critically ill patients with acute kidney injury. J Am Soc Nephrol 2006;17:3132–3138.

5 Bellomo R, Farmer M, Boyce N: The outcome of critically ill elderly patients with severe acute renal failure treated by continuous hemodiafiltration. Int J Artif Organs 1994;17:466–472.

6 Guerin C, Girard R, Selli JM, Ayzac L: Intermittent versus continuous renal replacement therapy for acute renal failure in intensive care units: results from a multicenter prospective epidemiological survey. Intensive Care Med 2002;28:1411–1418.

7 Lins RL, Elseviers MM, Van der Niepen P, Hoste E, et al: Intermittent versus continuous renal replacement therapy for acute kidney injury patients admitted to the intensive care unit: results of a randomized clinical trial. Nephrol Dial Transplant 2009;24:512–518.

8 Bellomo R, Boyce N: Continuous veno-venous hemodiafiltration compared with conventional dialysis in critically ill patients with acute renal failure. ASAIO J 1993;39:M794–M797.

9 Investigators RRTS, Bellomo R, Cass A, Cole L, et al: Intensity of continuous renal-replacement therapy in critically ill patients. N Engl J Med 2009;361:1627–1638.

10 Tolwani AJ, Campbell RC, Stofan BS, Lai KR, et al: Standard versus high-dose CVVHDF for ICU-related acute renal failure. J Am Soc Nephrol 2008;19:1233–1238.

11 Leblanc M, Thibeault Y, Querin S: Continuous haemofiltration and haemodiafiltration for acute renal failure in severely burned patients. Burns 1997;23:160–165.
12 Kanno Y, Nemoto H, Nakamoto H, Okada H, et al: Selection of hemoperfusion therapy for patients with septic shock on the basis of the primary disease. J Artif Organs 2003;6: 205–210.
13 John S, Griesbach D, Baumgartel M, Weihprecht H, et al: Effects of continuous haemofiltration vs. intermittent haemodialysis on systemic haemodynamics and splanchnic regional perfusion in septic shock patients: a prospective, randomized clinical trial. Nephrol Dial Transplant 2001;16:320–327.
14 Davenport A, Will EJ, Davidson AM: Improved cardiovascular stability during continuous modes of renal replacement therapy in critically ill patients with acute hepatic and renal failure. Crit Care Med 1993;21:328–338.
15 Baldwin I, Bellomo R, Naka T, Koch B, et al: A pilot randomized controlled comparison of extended daily dialysis with filtration and continuous veno-venous hemofiltration: fluid removal and hemodynamics. Int J Artif Organs 2007;30:1083–1089.
16 Bellomo R, Farmer M, Bhonagiri S, Porceddu S, et al: Changing acute renal failure treatment from intermittent hemodialysis to continuous hemofiltration: impact on azotemic control. Int J Artif Organs 1999;22:145–150.
17 Clark WR, Mueller BA, Alaka KJ, Macias WL: A comparison of metabolic control by continuous and intermittent therapies in acute renal failure. J Am Soc Nephrol 1994;4:1413–1420.
18 Barenbrock M, Hausberg M, Matzkies F, de la Motte S, et al: Effects of bicarbonate- and lactate-buffered replacement fluids on cardiovascular outcome in CVVH patients. Kidney Int 2000;58:1751–1757.
19 VA/NIH Acute Renal Failure Trial Network, Palevsky PM, Zhang JH, O'Connor TZ, et al: Intensity of renal support in critically ill patients with acute kidney injury. N Engl J Med 2008;359:7–20.
20 Rabindranath K, Adams J, Macleod AM, Muirhead N: Intermittent versus continuous renal replacement therapy for acute renal failure in adults. Cochrane Database Syst Rev 2007:CD003773.
21 Uehlinger DE, Jakob SM, Ferrari P, Eichelberger M, et al: Comparison of continuous and intermittent renal replacement therapy for acute renal failure. Nephrol Dial Transplant 2005;20:1630–1637.
22 Uchino S, Bellomo R, Kellum JA, Morimatsu H, et al: Patient and kidney survival by dialysis modality in critically ill patients with acute kidney injury. Int J Artif Organs 2007;30:281–292.
23 Vinsonneau C, Camus C, Combes A, Costa de Beauregard MA, et al: Continuous venovenous haemodiafiltration versus intermittent haemodialysis for acute renal failure in patients with multiple-organ dysfunction syndrome: a multicentre randomised trial. Lancet 2006;368:379–385.

Hiromichi Suzuki
Department of Nephrology, School of Medicine, Faculty of Medicine, Saitama Medical University
38 Morohongo Moroyama, Iruma
Saitama, 350-0495 (Japan)
Tel. +81 49 276 1620, Fax +81 49 295 7338, E-Mail iromichi@saitama-med.ac.jp

Controversies in Blood Purification in Critical Care

Suzuki H, Hirasawa H (eds): Acute Blood Purification.
Contrib Nephrol. Basel, Karger, 2010, vol 166, pp 173–180

Online CHDF System: Excellent Cost-Effectiveness for Continuous Renal Replacement Therapy with High Efficacy and Individualization

M. Takatori[a] · M. Yamaoka[a] · S. Nogami[a] · K. Ojima[b] · T. Masuda[b] · S. Takeuchi[b] · K. Tada[a]

Departments of [a]Anesthesia and Intensive Care and [b]Clinical Engineering, Hiroshima City Hospital, Hiroshima, Japan

Abstract

We developed an online continuous hemodiafiltration (CHDF) system with a central reverse osmosis (RO) fluid delivery system in 1996. This was improved to a system composed of a single-patient dialysis machine and RO module in 2003. This comprises a water treatment system, an RO module, a dialysis machine with 3 endotoxin retentive filters, 2 additional roller pump units, and a disposable special circuit. Dialysate is produced online by a dialysis machine using RO water and dialysate concentrate, which passes through endotoxin retentive filters and is supplied via the machine in the usual manner. A disposable special circuit and additional two roller pumps independently regulate dialysate flow and substitute flow from 0 to 12 in steps of 0.1 l/h. Seventy-seven patients with acute kidney injury (AKI) were treated with online CHDF from December 1996 to June 2004. Patient outcome was compared with the other modality of continuous renal replacement therapy from July 1992 to June 2004. The survival rates of each modality were 68.3, 65.0, 56.6 and 74.0% for conventional CHDF, high-flow continuous hemodialysis, high-flow CHDF and high-flow/high-volume CHDF (online CHDF), respectively. The survival rate of the high-volume modality (online CHDF) group was significantly higher ($p = 0.046$) than that of the low volume modality group (61.1%). Increases in efficacy and efficiency are a challenge facing blood purification therapy, and, moreover, individualized prescriptions are crucial in AKI patients in ICU. However, the cost of the dialysate and substitution fluid is a limitation of the therapy. The greatest advantage of the system is that a very high dose of delivered dialysate and substitute does not lead to a proportional rise in the cost. The online CHDF system is currently one of the most feasible solutions.

The optimal efficacy of continuous renal replacement therapy (CRRT) for critically ill patients with acute kidney injury (AKI) is still controversial. Although some recent trials [1, 2] have demonstrated that the dose of renal replacement therapy does not change the mortality of those patients, evidence from many well-designed randomized controlled studies still has to be considered [3, 4]. Moreover, the optimization of the efficacy of CRRT to each patient achieved by measuring the patient's metabolic status may be beneficial.

In this article, we report on the development, equipment construction and management of the online continuous hemodiafiltration (CHDF) system, and then discuss its usefulness and cost-effectiveness.

Methods and Results

Development of the Online CHDF System

We developed a high-flow CHDF system with a central reverse osmosis (RO) fluid delivery system and a dialysis machine in 1992. From 1992 to 1996, high efficacy CHDF was performed with this system. During this period, dialysate was supplied by the dialysis machine and Sublood B (Fuso, Japan) was used as substitution fluid. In 1996, after establishing an online supply of endotoxin-free substitution fluid from the dialysis machine, the clinical use of online CHDF system was started. This system was improved to a system composed of a single-patient dialysis machine and an RO module in 2003.

Equipment Comprising the Online CHDF System

The online CHDF system is composed of a water treatment system, an RO module (MH500CX, JWS, Japan), a dialysis machine (DBB-26, NIKKISO, Japan) with 3 endotoxin retentive filters (ETRF), 2 additional roller pump units, and a disposable special circuit. The dialysis machine's hardware and software were not modified in any way (fig. 1).

Dialysate produced online by the dialysis machine from RO water and dialysate concentrate passes through ETRFs and is supplied from the machine in the usual manner. The dialysate flow rate is usually set to about 300 ml/min (18,000 ml/h), which is branched to dialysate, substitute and the 'bypass' flows. A disposable special circuit for the online CHDF system is connected to the dialysate supply and return ports. Additional 2 roller pumps, one to control the dialysate flow and another to control the substitute flow, are placed on the dialysis machine and their slave control cables are connected to the monitor output. These roller pumps can independently regulate each flow from 0 to 12 in steps of 0.1 l/h. Residual flow from the dialysate supply port of the machine passes a 'bypass circuit' of a special circuit to the return port without contact with the patient's blood.

The in/out balance of the dialysate is strictly controlled by the dialysis machine. Two roller pumps may have considerable flow errors, but in/out balance depends only on the balancing system of the dialysis machine. Consequently, the filtration rate via the filter is equal to the substitute flow rate. This balancer system with duplex pump is comparable or even superior to that of conventional CHDF machines.

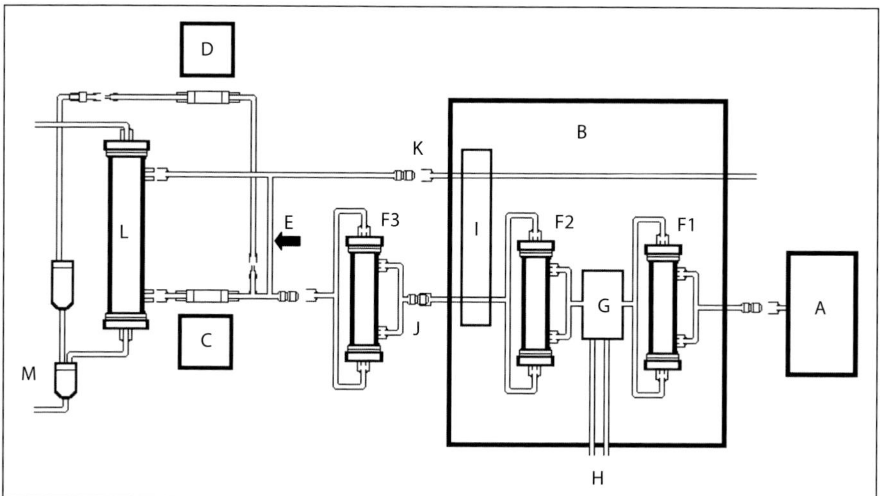

Fig. 1. Equipment comprising the online CHDF system. A = RO module (MH500CX JWS); B = dialysis machine (DBB-26 NIKKISO); C = roller pump to control dialysate flow; D = roller pump to control substitution flow; E = disposable special circuit for the online CHDF (arrow indicates bypass circuit); F1–F3 = ETRF; G = mixer; H = dialysate concentrate; I = duplex pump; J = dialysate supply port; K = dialysate return port; L = dialyzer; M = venous drip chamber of blood circuit.

Application to Patients with AKI

Patients were treated using the online CHDF system with a blood flow rate of 100 ml/min, dialysate flow rate of 1,200–10,000 ml/h (mostly 2,400 ml/h), and substitute flow rate of 1,000–3,000 ml/h (mostly 1,500 ml/h). The filter types used were PS-1.6UW (Kawasumi, Japan) and FX-140 (Fresenius Medical Care, Japan). The bicarbonate concentration of the dialysate was usually set to 28 mEq/l, and was adjusted (upper limit of about 70 mEq/l) according to the patient's acidosis. The dialysate and substitute flow rates were adjusted for each patient according to the assessment of hemodynamic stability, temperature control, nutritional requirement, nitrogen balance and blood urea level. In hemodynamically unstable patients with severe azotemia, a gradual increment of the flow rate was programmed to avoid inter-space osmolality differences and to maintain the intravascular blood volume.

Maintenance

The acid-cleaning using acetic acid and the chemical disinfection using sodium hypochlorite are performed after every 48 h of continuous use for CRRT and after the completion of the treatment. When the system is out of use, it is connected to the piping in the preparation room and is rinsed 3 times every week, disinfected twice a week and acid-cleaned once a week. The process is programmed automatically.

Quality Control for Sterility

From 1996, the endotoxin levels of the dialysate were measured just before the start of each treatment, and were validated to be less than the detection limit (<1 IU/l) before

Table 1. The modality of CRRT and patient outcome

Group	Modality	n	Median dialysate flow, ml/h	Median substitute flow, ml/h	SOFA score (mean ± SD)	Survival, %
1	conventional CHDF	41	400	400	10.29±3.37	68.3
2	high-flow CHD	40	3,000	0	11.00±3.26	65.0
3	high-flow CHDF	99	3,000	400	12.01±3.59	56.6
4	high-flow/high-volume CHDF	77	2,400	1,500	10.72±3.76	74.0

and after the final ETRF (F3 in fig. 1). After this quality was validated, monthly endotoxin monitoring was continued. ETRFs are exchanged every 3 months.

Patient Outcome

Seventy-seven patients with AKI were treated with online CHDF from December 1996 to June 2004. No side effects were noted. Azotemia and acid-base balance were efficiently controlled. Potassium and phosphate occasionally needed to be replaced. Hemodynamic parameters were evaluated with the Swan-Ganz catheter in a few cases, which showed a superior effect in hemodynamic improvement.

Patient outcome was compared with the other modality of CRRT. Two hundred fifty-seven patients with AKI treated with CRRT from July 1992 to June 2004 were retrospectively investigated. Statistical analysis was performed by the one-sample z-test.

The modality of CRRT was classified into 4 categories: conventional CHDF, high-flow continuous hemodialysis (CHD), high-flow CHDF, and high-flow/high-volume CHDF (online CHDF), as shown in table 1. Conventional CHDF (group 1, 41 patients) was delivered by a CHDF machine (Jun500, Ube, Japan) and Sublood B was used as dialysate and substitute. The median dialysate and substitute flow was 400ml/h each, which is restricted by the insurance limit in Japan. High-flow CHD (group 2, 40 patients) was delivered by dialysis machine. This system is similar to the online CHDF system, except for a substitute supply line. High-flow CHDF (group 3, 99 patients) was delivered by the same system, wherein Sublood B was used as substitute at a rate of 400 ml/h. High-flow/high-volume CHDF (online CHDF; group 4, 77 patients) has been described previously.

The pathological causes of AKI were quite variable: about 35% sepsis and 30% low output syndrome (LOS) and ischemic damage. The SOFA scores for each group (means ± SD) were 10.29 ± 3.37 (group 1), 11.00 ± 3.26 (group 2), 12.01 ± 3.59 (group 3) and 10.72 ± 3.76 (group 4), respectively (no statistical difference between each group). The total survival rate of all cases was 65.0%, and the survival in each group was 68.3% (group 1), 65.0% (group 2), 56.6% (group 3) and 74.0% (group 4), respectively. Only groups 3 and 4 exhibited a statistical difference ($p = 0.016$). Next, the high-volume modality (group 4) and low-volume modality groups (groups 1, 2 and 3) were compared, and the results are shown in figure 2. The survival rate of the

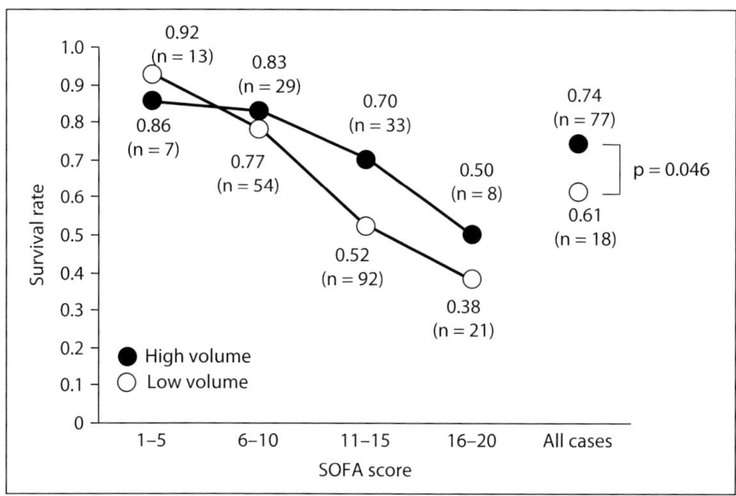

Fig 2. Comparison of the survival rate of high-volume modality group (i.e. online CHDF, n = 77) and low-volume modality group (i.e. all other groups, n = 180).

high-volume modality group was 74.0%, which was significantly higher ($p = 0.046$) than that of the low-volume modality group (61.1%).

Discussion

Efficacy of acid-base balance correction in online CHDF is calculated about 6 times as high as in conventional CHDF. Kindaly Solution AF-2 (Fuso, Japan), the dialysate used from 1992 to 2004, contains 8 mEq/l of acetate. The small amount of acetate in bicarbonate dialysis fluid may allow significant transfer of that anion to the patient, possibly inducing cytokine activation [5]. In 2007, a new dialysate Carbostar-L (Ajinomoto, Japan), which contains citrate and no acetate, became available in Japan. This has been used for high-flow CHD in our ICU since 2008. Acetate-free biofiltration, based on the technique of separately infusing bicarbonate (administered after diafiltration), is one of the alternative ways. This method has the advantages of controlling metabolic acidosis and cardiovascular instability [6]. Online CHDF with Carbostar-L is considered to have the same advantages.

Deposition of calcium base is a major problem in the system. To avoid this, periodic acid cleaning is necessary every 48 h of continuous running. During the several hours of acid cleaning, CRRT must be discontinued. Another machine is prepared for continuing the therapy if the patient's hemodynamic status is unstable at the time of discontinuation of CRRT. Bicarbonate-free dialysate may provide stable long-time running of dialysis machines without deposition of calcium in the lines.

In most cases of AKI patients in the ICU, negative nitrogen balance and enhanced production of urea with sustained accelerated protein breakdown are the key characteristics. In those cases, although the catabolic status appears unresponsive to artificial nutrition providing exogenous nutritional substrates, nitrogen balance is still shown to be positively related to protein intake. Higher protein/amino acid intake is associated with positive (or less negative) nitrogen balance and an overall increase in protein catabolic rate, and then the dose of renal replacement therapy should be increased accordingly [7]. The protein requirement in patients with AKI and severe hypercatabolism has been shown to be 1.5–2.0 g/kg/day [7]. Recently, the term 'protein-energy wasting' has been proposed, and hyponutrition was shown to be a risk factor of mortality in either CKD or AKI patients [8]. In clinical settings, we can easily measure urea appearance and adjust protein intake, and finally adjust CRRT dose for increased urea production with the online CHDF system. Although no data are available from randomized controlled trials on the optimal nutrition and renal replacement therapy dose, the daily Kt/V required to keep urea balance reaches about 1.4–2.5 in our cases. In septic AKI, the optimal dose of CRRT is less clear. The mechanisms behind the beneficial effects of CRRT may be complex. Although the details are beyond the scope of this paper, we would like to emphasize the direct cell immunomodulatory action during renal replacement therapy [9]. Using online CHDF, we sometimes saw good regulation of body temperature and a marked decrease in circulating IL-6 without a reduction in the concentration by the filter. CRRT dose is directly related to the blood temperature, temperature in the circuit as well as body temperature, and it may alter the immune cell function.

As in online hemofiltration, patients are directly infused with online substitution fluid in online CHDF. So, adequate water treatment and quality control are essential. In 1996, at the start of online CHDF, we validated the water quality according to the standards proposed by the Kyushu Society for HDF [10], which have been used as the basis for all subsequent standards. Recently, a new quality standard for dialysis water was proposed, but a standard method to measure microbiological contaminants has not been established. Kawanishi et al. [11], the Committee of Scientific Academy of Japanese Society for Dialysis Therapy, proposed a new standard on microbiological management of fluids. The recommendation shows a concrete and practical method to validate the quality. In the ICU, however, systems are used irregularly. Machines are sometimes disconnected from lines and moved from patient to patient, and sometimes are not used for extended periods. In such circumstances, validation should be confirmed in each facility. Our experimental data show that the endotoxin level is kept low during continuous running (table 2), but it increases notably while the system is not in use, especially in the line between the RO module and dialysis

Table 2. Endotoxin level of RO water and dialysate under CRRT (IU/l)

Time after start, h	Before prefilter[1] (RO)	Before first filter[2] (D1)	After final filter[3] (D2)
0	3.221	<0.614	<0.614
6	4.834	<0.614	<0.614
12	1.971	<0.614	<0.614
24	3.969	<0.614	<0.614
48	3.634	<0.614	<0.614

[1] Port between A and F1 (fig. 1).
[2] Port between G and F2 (fig. 1).
[3] After F3 (fig. 1).

machine. So, the periodic cleaning of the system while it is not in use is indispensable. Quality control, which has previously been validated by measuring the endotoxin level, should be re-validated in accordance with the new standard.

After Ronco et al. [4] demonstrated the superiority of high filtration ratio, only 11.7% of the patients were treated with a dose >35 ml/kg/h [12]. In many centers in the world, as in Japan, the cost of substitution fluid is the major restrictive factor of high efficacy CRRT. Thus, the daily costs for substitution fluid were compared. Sublood BS (Fuso, Japan), which is most often used in Japan as substitution fluid, entails a daily cost of JPY 28,680 at a rate of 2,000 ml/h, and JPY 86,040 at a rate of 6,000 ml/h. On the other hand, when acetate-free dialysate Carbostar-L is used with the online CHDF system, it costs only JPY 5,186 regardless of dialysate and substitute flow rates. Additional costs for the maintenance of the water treatment system may be estimated at about JPY 2,000 for 1 treatment session. Thus, the online CHDF system is the modality with the best cost-effectiveness when high efficacy and individualized CRRT are desired.

Conclusion

Increases in efficacy and efficiency are a challenge for blood purification therapy, and, moreover, individualized prescriptions are crucial for AKI patients in the ICU. Even though the cost of dialysate and substitution fluid is the main limitation of the therapy, the online CHDF system that was developed and described here may be one of the most feasible solutions.

References

1 Vesconi S, Cruz DN, Fumagalli R, Kindgen-Milles D, Monti G, Marinho A, Mariano F, Formica M, Marchesi M, René R, Livigni S, Ronco C: Delivered dose of renal replacement therapy and mortality in critically ill patients with acute kidney injury. Crit Care 2009;13:R57.

2 Ronco C, Cruz D, Oudemans van Straaten H, Honore P, House A, Bin D, Gibney N: Dialysis dose in acute kidney injury: no time for therapeutic nihilism – a critical appraisal of the Acute Renal Failure Trial Network study. Crit Care 2008;12:308–314.

3 Saudan P, Niederberger M, Seigneux SD, Romand J, Pugin J, Perneger T, Martin PY: Adding a dialysis dose to continuous hemofiltration increases survival in patients with acute renal failure. Kidney Int 2006;70: 1312–1317.

4 Ronco C, Bellomo R, Homel P, Brendolan A, Dan M, Piccinni P, La Greca G: Effects of different doses in continuous veno-venous haemofiltration on outcomes of acute renal failure: a prospective randomised trial. Lancet 2000;356:26–30.

5 Pizzarelli F, Cerrai T, Dattolo P, Ferro G: On-line haemodiafiltration with and without acetate. Nephrol Dial Transplant 2006;21: 1648–1651.

6 Galli GP: Acetate free biofiltration (AFB): from theory to clinical results. Clin Nephrol 1998;50:28–37.

7 Fiaccadori E, Parenti E, Maggiore U: Nutritional support in acute kidney injury. J Nephrol 2008;21:645–656.

8 Fouque D, Kalantar-Zadeh K, Kopple J, Cano N, Chauveau P, Cuppari L, Franch H, Guarnieri G, Ikizler TA, Kaysen G, Lindholm B, Massy Z, Mitch W, Pineda E, Stenvinkel P, Treviño-Becerra A, Wanner C: A proposed nomenclature and diagnostic criteria for protein-energy wasting in acute and chronic kidney disease. Kidney Int 2008;73:391–398.

9 Girndt M, Kaul H, Leitnaker CK, Sester M, Sester U, Köhler H: Selective sequestration of cytokine-producing monocytes during hemodialysis treatment. Am J Kidney Dis 2001;37:954–963.

10 Sato T, Takamiya T, Kim ST, Yamomoto C, Fukui H, Nakamoto M: Dialysate and substitution fluid quality for online haemodiafiltration and hemofiltration. Nephrology 1997;3:549–555.

11 Kawanishi H, Akiba T, Masakane I, Tomo T, Mineshima M, Kawasaki T, Hirakata H, Akizawa T: Standard on microbiological management of fluids for hemodialysis and related therapies by the Japanese Society for Dialysis Therapy. Ther Apher Dial 2009;13: 161–166.

12 Uchino S, Bellomo R, Morimatsu H, Morgera S, Schetz M, Tan I, Bouman C, Macedo E, Gibney N, Tolwani A, Oudemans-van Straaten H, Ronco C, Kellum JA: Continuous renal replacement therapy: a worldwide practice survey. The beginning and ending supportive therapy for the kidney (B.E.S.T. kidney) Investigators. Intensive Care Med 2007;33:1563–1570.

M. Takatori
Department of Anesthesia and Intensive Care, Hiroshima City Hospital
7–33 Motomachi, Nakaku
Hiroshima City, Hiroshima 730-8518 (Japan)
Tel. +81 82 221 2291, Fax +81 82 223 1447, E-Mail takatori@eagle.ocn.ne.jp

Suzuki H, Hirasawa H (eds): Acute Blood Purification.
Contrib Nephrol. Basel, Karger, 2010, vol 166, pp 181–189

'Super High-Flux' or 'High Cut-Off' Hemofiltration and Hemodialysis

Toshio Naka[a] · Michael Haase[b] · Rinaldo Bellomo[c]

[a]Department of Critical Care Medicine, Wakayama Medical University, Wakayama, Japan;
[b]Department of Nephrology and Intensive Care, Charite University Medicine, Berlin, Germany;
[c]Department of Intensive Care, Austin Hospital, University of Melbourne, Melbourne, Australia

Abstract

Recently, 'super high-flux' (SHF) or 'high cut-off' (HCO) membranes have been developed to increase the clearance of inflammatory mediators. In the experimental and clinical settings, SHF/HCO membranes appear to achieve greater clearance of inflammatory cytokines than conventional high-flux membranes. SHF/HCO membranes also restore immune cell function, attenuate hemodynamic instability and decrease plasma IL-6 levels. Moreover, SHF/HCO membranes can eliminate larger late-phase inflammatory mediators such as HMGB-1. Although albumin sieving coefficients with SHF/HCO membranes are greater than with conventional high-flux membranes, the daily amount lost is limited and can be replaced. Hemodialysis with SHF/HCO membranes can also achieve similar cytokine removal to hemofiltration with acceptable albumin losses. When strategies for sepsis or systemic inflammation treatment target middle molecular mediators, both SHF/HCO hemofiltration and hemodialysis appear feasible and safe and require further clinical investigation.

The mortality rate of severe sepsis associated with acute kidney injury (AKI) remains high (approximately 60%) despite advances in our knowledge of its pathophysiology [1, 2]. Humoral mediators such as inflammatory cytokines play a major role in the pathophysiology of systemic inflammatory response syndrome (SIRS) or sepsis [3, 4]. In the last 2 decades, various extracorporeal blood purification methods have been used in the treatment of SIRS/sepsis to increase the clearance capacity for inflammatory mediators. However, substantial elimination of these mediators has not been achieved with conventional membranes, in part because of the relatively large molecular weight of septic middle molecules. Accordingly, a new approach is needed to remove middle-molecular-weight inflammatory mediators with extracorporeal renal replacement therapy.

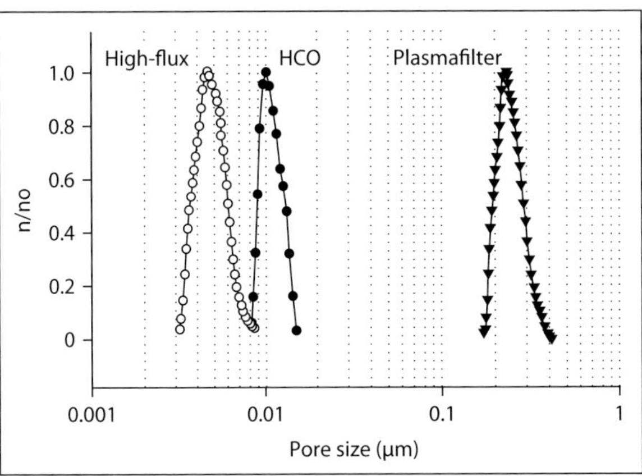

Fig. 1. Pore sizes of SHF or HCO membranes in comparison to conventional high-flux and plasmafiltration membranes. n/no = Relation (quotient) of the number of pores of a certain size (n) vs. the number of pores with the commonest size (no) (peak).

Recently, 'super high-flux' (SHF) or 'high cut-off' (HCO) hemofiltration (HF) and hemodialysis (HD) have been introduced to overcome the limitations seen with small pore size. This article aims to explore the experimental and clinical effects of such treatment modalities.

SHF/HCO Hemofiltration

Inflammatory cytokines such as TNF-α, IL-6 and IL-8 play a major role in the pathophysiology of SIRS/sepsis [3, 4]. Apart from creatinine (MW: 113 Da) or other low-molecular-weight substances, these mediators are relatively large (TNF-α: 17 kDa, IL-6: 26 kDa and IL-8: 8 kDa), and are classified as middle molecular substances. These substances cannot pass freely through the pores of conventional high-flux membranes, and therefore their clearances are limited. To overcome this limitation, a new approach has been introduced by increasing the pore size of membranes in an attempt to improve the clearance of inflammatory mediators. These membranes have been called 'super high-flux' membranes by Uchino et al. [5] and 'high cut-off' point membranes by Morgera et al. [6]. Several SHF/HCO membranes have been manufactured so far. These membranes are made from polyarylethersulfone, polysulfone or cellulose triacetate [7]. Although the definition of SHF/HCO membrane has not been strictly established, pore sizes are around 0.01 μm (fig. 1) [7]. Therefore, the pore sizes of SHF/HCO membranes are 2- or 3-fold larger than conventional high-flux membranes, which have a pore size of 0.003–0.006 μm, and one twentieth of plasma-filtering membranes which have which have a pore size of around 0.2

Table 1. SCs (%) of SHF/HCO and conventional pore membranes for cytokines and albumin

Membrane	First author	IL-8 (8 kDa)	TNF-α (17 kDa)	IL-1β (17 kDa)	IL-6 (26 kDa)	Albumin (66 kDa)
100-kDa polyamide	Uchino [5]	31	27	81	73	6
100-kDa polyamide	Morgera [16]				87	4.8
150-kDa polyamide	Morgera [6]		78		99	5
80-kDa cellulose triacetate	Bordont [17]		72	58		6
60-kDa cellulose triacetate	Uchino [12]	24	32	62	54	6
50-kDa polyamide	Hoffmann [9]	25	0	18	0	
30-kDa polysulfone	Heering [10]	12	22	42	4	
30-kDa AN69	DeVriese [11]	8	16	22	18	

μm (fig. 1) [7]. The nominal cut-off points for SHF/HCO membranes range from 60 to 150 kDa and the clinical cut-off points in blood range from 40 to 100 kDa [7]. Thus, SHF/HCO membranes have the potential to increase the sieving coefficients (SCs) of various inflammatory cytokines, but also of albumin (MW: 66 kDa), AT-III (MW: 60 kDa), protein C (MW: 62 kDa) and others.

Ex vivo Findings with SHF/HCO Membranes

SCs for various cytokines and albumin have been investigated with conventional high-flux and SHF/HCO hemofilters (table 1) [8]. For conventional hemofilters, Hoffmann et al. [9] studied a polyamide membrane and reported SCs for IL-6 of 0% and IL-8 of 25%. Heering et al. [10] also used polysulfone membrane and reported an SC for IL-6 of 4% and IL-8 of 12%. Moreover DeVriese et al. [11] tested the AN69 membrane and reported an SC for IL-6 of 18% and IL-8 of 8%. The SCs with conventional high-flux hemofilters ranged from 5 to 20%, and sieving of cytokines is actually limited. This limitation resulted in the failure to decrease plasma cytokine concentration with renal replacement therapy.

In contrast, in 2002, Uchino et al. [12] performed an ex vivo experiment using a cellulose triacetate membrane, which has a nominal cut-off point of 60 kDa, and found higher SCs for IL-1β (62%), IL-6 (54%), IL-8 (24%), IL-10 (67%) and TNF-α (32%). Subsequently, in 2003, Uchino et al. [5] also assessed the newly developed SHF membrane derived from polyamide with a nominal cut-off point of 100 kDa. High SCs for cytokines were achieved: IL-1β (65%), IL-6 (67%), IL-8 (31%), IL-10 (65%) and TNF-α (28%). Morgera et al. [6] also tested a HCO membrane with a nominal cut-off point of 150 kDa. This HCO membrane resulted in high SCs for cytokines: IL-6 (50–109%), TNF-α (43–84%). These data consistently demonstrate that by increasing pore size SHF/HCO membranes could achieve 3- or 4- fold greater SCs, and therefore clearance, than conventional high-flux membranes.

Clearances for solutes with SHF/HCO membranes depend not only on SCs, but also blood flow and filtration/dialysate flow rate. In studies using continuous veno-venous HF (CVVH) or continuous veno-venous HD (CVVHD), blood flow was set at 150 ml/min and filtration/dialysate flow ranged from 17 to 42 ml/min [7]. Maximum IL-6 clearance for SFH/HCO-CVVH was 40 ml/min [13] and maximum IL-1ra clearance was 42 ml/min [14]. Moreover, SHF/HCO-HD achieved higher clearances when blood flow was 300 ml/min and dialysate flow rate was 500 ml/min. IL-6 clearance for SFH/HCO-HD was 67 ml/min, and IL-8 clearance was 62 ml/min [15].

Pilot Clinical Study of SHF/HCO Hemofiltration

With the support of in vitro data, Morgera et al. [16] carried out a single-center pilot clinical study in 2003. Sixteen patients with multiple organ failure secondary to septic shock were treated by HCO-HF for 12 h per day first, and alternated with conventional HF for 12 h over 5 days. The cut-off point for the HCO membrane was 60 kDa. Blood flow was set at 150ml/min and replacement volume was 1,000 ml/h (16.7 ml/min). HCO-HF was well tolerated. The SC for IL-6 was 0.87 (0.76–1.09) and remained stable throughout the 12-hour treatment. IL-6 clearances were 12–17 ml/min. Plasma IL-6 tended to decrease, but did not reach statistical significance. Subsequently, Morgera et al. [13] conducted a prospective randomized clinical trial. Thirty patients with septic AKI were allocated to either HCO-CVVH or conventional CVVH. Median replacement rate was 31 ml/kg/h. HCO-CVVH was associated with a significant reduction in the norepinephrine dose to maintain mean arterial pressure to 60 mm Hg (from 0.27 μg/kg/min at baseline to 0.07 μg/kg/min at the end of treatment; $p = 0.0002$). SCs for IL-6 with the HCO membrane were 0.84–0.93 and HCO-CVVH achieved a high clearance rate for IL-6 (36–40 ml/min). Subsequently, plasma IL-6 reduced significantly for HCO-CVVH over time ($p = 0.0465$), whereas values remained unchanged within the control group.

Effect of SHF/HCO Membrane on Cellular Function

Several investigators have studied the effect of SHF/HCO membranes on cellular function. Bordoni et al. [17] evaluated the effect of SHF/HCO membranes made from cellulose triacetate on apoptotic activity in an in vitro blood circulation model. Human whole blood spiked with *E. coli* was re-circulated with SHF/HCO or standard high-flux membrane. Incubation of SHF/HCO ultrafiltrate with U937 monocytes induced caspase-3 and caspase-8 activity on monocytes. Morgera et al. [18] studied the effect of HCO-HF on peripheral mononuclear cells (PBMC) function in a randomized controlled trial (RCT). Twenty-eight septic patients were randomly assigned to either HCO-HF or conventional HF. PBMC in septic patients reduced T lymphocyte proliferation. HCO-HF restored PBMC proliferation back into the normal range, whereas conventional HF did not. Incubation of HCO-HF ultrafiltrate from septic patients with PBMC from healthy volunteers resulted in a significant suppression of such proliferation, while standard HF ultrafiltrate did not. In addition, incubation of HCO ultrafiltrate from septic patients with PBMC from healthy volunteers resulted in a significant TNF release ($p < 0.001$). The same working group also assessed HCO-HF on polymorphonuclear leukocytes (PML) function [19]. The phagocytosis activity almost doubled compared to that of healthy volunteers. HCO-HF significantly decreased PML phagocytosis activity ($p < 0.05$), whereas standard HF did not. When HCO ultrafiltrate of septic patients incubated with blood from healthy donors, PML phagocytosis activity increased significantly ($p < 0.001$).

These observations demonstrate that HCO-HF exhibits immunomodulatory effects including inhibitory effects of apoptosis, restoration of PBMC proliferation, and a decrease in PML phagocytosis activity.

Albumin Loss and Safety Issues

SHF/HCO-HF might cause a loss of protein, especially albumin. Excessive albumin loss could be a clinical issue and an important limitation of these treatments.

Uchino et al. [5] reported that with SHF-HF, the SC of albumin was 6%. Morgera et al. [16] also assessed HCO-CVVH. The SC of albumin was 4.8% at 30 min; however, it decreased over time (2.0% at 1 h and 1.0% at 12 h). They suggested that this phenomenon could be due to blood component deposition in the pores of the membrane and subsequent narrowing of effective pore size. The cumulative albumin loss was 4.82 g. Subsequently, Morgera et al. [20] compared albumin loss of HCO-CVVH with that of HCO-CVVHD. Albumin clearances ranged from 0.06 ml/min (CVVHD at 2.5 l/h) to 1.7 ml/min (CVVH at 2.5 l/h), whereas clearances for IL-6 were similar. Albumin loss during CVVH or CHHVD was estimated between 5 and 10 g per day. This loss of albumin could be supplemented with 50 ml of 20% human albumin solution. In addition, plasma AT-III (60 kDa), factor II (69 kDa), VIII (265 kDa), protein C (62 kDa) and protein S (60 kDa) did not change during the HCO-CVVH treatment [16]. If cytokines can be removed effectively by diffusion, diffusive clearance has an advantage over convective clearance because of lower albumin loss.

HCO Hemodialysis

Cytokine clearance by diffusion has been tested by several investigators. Uchino et al. [21] studied SCs and clearance for cytokines with SHF-HD. Whole blood spiked with *E. coli* was re-circulated and the circuit attached to an SHF membrane. Blood flow was set at 250 ml/min, and dialysate flow rate was set at 1 and 9 l/h, then investigators assessed the dialysate/plasma ratio (D/P) and clearance for cytokines and albumin. HCO-HD achieved high D/P ratio for cytokine (IL-6: 0.67, IL-8: 0.94, and albumin: 0.11) at 1 l/h dialysate flow. D/P ratio decreased when dialysate increased to 9 l/h; however, clearances for cytokines increased (IL-6: 19 ml/min, IL-8: 51 ml/min). Clearance for albumin was 2.4 ml/min at 1 l/h dialysate, and decreased to 1.2 ml/min at 9 l/h. Lee et al. [22, 23] studied cytokine and β_2-microglobulin (β_2M; MW: 11.8 kDa) removal with SHF-HD in an ex vivo study. Blood flow was set at 300 ml/min, and dialysate flow rate was set 200, 300 and 500 ml/min. SHF-HD achieved high cytokine and β_2M clearances (IL-6: 66.8 ml/min, IL-8: 61.7 ml/min, TNF-α: 36.1 ml/min, β_2M: 113.5 ml/min). Cytokine and β_2M clearance did not increase with increasing the dialysate flow to 500 ml/min. Albumin clearance was 2.7 ml/min at 200 ml/min of dialysate flow and 5.4 ml/min at 500 ml/min of dialysate flow. Lee et al. [24] also evaluated β_2M removal with HCO-HD in a randomized double-blind crossover study. Eight stable HD patients were randomly assigned to HCO-HD and high-flux membrane HD for 2 weeks each, between a 1-week washout period. HCO-HD resulted in significantly lower post-dialysis β_2M (10.8 vs. 14.2 mg/l, $p = 0.003$) than high-flux HD. HCO-HD was associated with a decrease in albumin (from 36 to 29.5 g/l, $p = 0.018$); however, the albumin level returned to baseline after a washout period (to 33.5 g/l). Clearance of albumin with HCO-HD was 2.2 ml/min. Moreover, Haase et al. [25] conducted a double-blind crossover RCT. Ten septic patients with AKI (RIFLE class F) were assigned to either 4 h of HCO-IHD or high-flux HD, following the other arm [25]. HCO-HD was associated with a greater decrease in plasma IL-6 (30.3%) than high-flux HD (1.1%, $p = 0.04$). Albumin loss with HCO-HD was 7.7 g.

These observations demonstrated that cytokines could be actually removed by diffusion with acceptable albumin loss and justified further RCT. An RCT for HCO-CVVHD is currently ongoing in Austria and Germany, and named the High Cut-Off CVVHD in Patients Treated for Acute Renal Failure after SIRS/Septic Shock (HICOSS) study (http://clinicaltrials.gov/ct2/show/NCT00875888). The HICOSS study is a prospective double-blind multicenter RCT; 120 patients treated for AKI after SIRS/septic shock requiring catecholamine administration will be randomly assigned to either HCO-CVVHD or conventional high-flux membrane CVVHD. Blood flow will be set at 200 ml/min and dialysate flow rate at 35 ml/kg/h. Primary outcomes are dosages of vasopressors, mean arterial pressure, heart rate and central venous pressure. The estimated study completion date is February 2010.

Table 2. Sieving coefficients (%) of super high-flux and conventional high-flux membranes for cytokines, HMGB-1 and albumin

	HMGB-1 (MW: 30 kDa)	IL-6 (MW: 26 kDa)	IL-8 (MW: 8 kDa)	Albumin (MW: 66 kDa)
High-flux				
1 l/h	1.2 (0.9, 4.5)	5.6 (4.9, 7.1)	10.8 (8.3, 13.8)	0.3 (0.2, 0.3)
Super high-flux				
1 l/h	15.5 (13.1, 19.0)*	79.5 (70.5, 79.6)*	68.1 (61.2, 73.4)*	2.5 (1.9, 3.6)*
2 /h	12.2 (10.1, 14.6)*	79.8 (69.0, 80.1)*	63.4 (62.7, 68.9)*	1.2 (1.1, 1.4)*

Data are shown as medians (25th, 75th percentile). * $p < 0.05$.

Another Therapeutic Target: HMGB-1 and SHF/HCO Membrane

Evidence has accumulated which indicates that the SHF/HCO membrane can eliminate inflammatory cytokines with high SCs and reduce plasma cytokine levels. However, whether the SHF/HCO membrane has a beneficial effect on physiology or survival remains unknown. Of course, not only inflammatory cytokines but also many other mediators have a role in the pathophysiology of SIRS/sepsis. Recently, Parrish et al. [26, 27] proposed that inflammatory mediators which appear in the late phase as may be a target of therapeutic strategies. High-mobility group box-1 (HMGB-1) is one of these mediators. HMGB-1 was originally identified as a nuclear DNA-binding protein to stabilize DNA structure [27]. Wang et al. [28] found HMGB-1 acts as a 'late' inflammatory cytokine that contributes to the progression of sepsis and other inflammatory responses. In a mouse model, HMGB-1 was induced in blood 8–32 h after endotoxin exposure [28]. Delayed administration of antibodies to HMGB-1 attenuated endotoxin lethality and administration of HMGB-1 itself was lethal [28]. Thus, HMGB-1 is called a 'lethal' mediator. Increased HMGB-1 levels were also reported to contribute to acute lung injury [29]. HMGB-1 can be secreted by activated immune cells or passively released by injured or necrotic cells. The MW of HMGB-1 is approximately 30 kDa, and therefore larger than other early phase inflammatory cytokines, and its half-life is estimated to be more than 6 h [28]. Accordingly, HMGB-1 cannot be removed by conventional high-flux membranes. However, SHF/HCO membranes should remove it, which could have a clinical benefit. We recently tested the SC for HMGB-1 with an SHF/HCO membrane. Whole blood from healthy donors was spiked with *E. coli* and cultured for 8 h to induce both cytokines and HMGB-1, then re-circulated with either an SHF/HCO membrane or a high-flux membrane. The SC for HMGB-1 was 12.2–15% with SHF/HCO, while it was 1.2% with high-flux membranes (table 2). Estimated clearances for HMGB-1 were 2.6 ml/min at 1 l/h of filtrate and 4.1 ml/min at 2 l/h of filtrate. In contrast to much higher clearance rates for cytokines, clearance for HMGB-1 appeared to be low. However, the half-life of HMGB-1 is much longer (>6 h) than the half-lives of cytokines (5–10 min). Therefore, SHF/HCO might have the potential to attenuate plasma HMGB-1 levels despite smaller clearance rates. Further clinical investigation is needed to establish whether SHF/HCO membranes attenuate plasma HMGB-1 levels and have any clinical benefit.

Conclusions

In conclusion, SHF/HCO membranes appear to be well tolerated, and achieve greater clearance of inflammatory cytokines than conventional high-flux membranes. SHF/HCO membranes also restore immune cell function, attenuate hemodynamic instability and decrease plasma IL-6 levels. Although albumin is removed with SHF/HCO membranes more than conventional high-flux membranes, the amount is relatively small and could be supplemented. Evidence has accumulated through ex vivo experimental studies and in vivo clinical studies. When strategies for SIRS/sepsis treatment target middle molecular mediators, super high-flux/ high cut-off point HF and HD appear feasible and safe modalities, which require further clinical investigation.

References

1 Dellinger RP: Cardiovascular management of septic shock. Crit Care Med 2003;31:946–945.

2 Bagshaw SM, Laupland KB, Doig CJ, et al: Prognosis for long term survival and renal recovery in critically ill patients with severe acute renal failure: a population based study. Crit Care 2005;9:R700–R709.

3 Thijs LG, Hack CE: Time course of cytokine levels in sepsis. Intensive Care Med 1995;21: s258–s263.

4 Van Dissel JT, Van Langevelde P, Westendrop RG, et al: Anti-inflammatory cytokine profile and mortality in febrile patients. Lancet 1998;351:950–953.

5 Uchino S, Bellomo R, Goldsmith D, et al: Super high flux hemofiltration: a new technique for cytokine removal. Intensive Care Med 2002;28:651–655.

6 Morgera S, Klonower D, Rocktaschel J, et al: TNF-α elimination with high cut-off haemofilters: a feasible clinical modality for septic patients? Nephrol Dial Transplant 2003;18: 1361–1369.

7 Haase M, Bellomo R, Morgera S, et al: High cut-off point membranes in septic acute renal failure: a systematic review. Int J Artif Organs 2007;30:1031–1041.

8 Honore PM, Matson JR: Hemofiltration, adsorption, sieving and the challenge of sepsis therapy design. Crit Care 2002;6:394–396.

9 Hoffmann J, Werdan K, Hartl W, et al: Hemofiltration from patients with severe sepsis and depressed left ventricular contractility contains cardiotoxic compounds. Shock 1999;12:174–180.

10 Heering P, Morgera S, Schmitz F, et al: Cytokine removal and cardiovascular hemodynamics in septic patients with continuous veno-venous hemofiltration. Intensive Care Med 1997;23:288–296.
11 De Vriese A, Colardyn F, Philippe JJ, et al: Cytokine removal during continuous hemofiltration in septic patients. J Am Soc Nephrol 1999;10:846–853.
12 Uchino S, Bellomo R, Goldsmith D, et al: Cytokine removal with a large pore cellulose triacetate filter: an ex vivo study. Int J Artif Organs 2002;25:27–32.
13 Morgera S, Haase M, Kuss T, et al: Pilot study on the effects of high cutoff hemofiltration on the need for norepinephrine in septic patients with acute renal failure. Crit Care Med 2006;34:2099–2104.
14 Morgera S, Slowinski T, Melzer C, et al: Renal replacement therapy with high-cutoff hemofilters: impact of convection and diffusion on cytokine clearances and protein status. Am J Kidney Dis 2004;43:444–453.
15 Lee WC, Uchino S, Fealy N, et al: Super high flux hemodialysis at high dialysate flows: an ex vivo assessment. Int J Artif Organs 2004;27:24–28.
16 Morgera S, Rocktaschel J, Haase M, et al: Intermittent high permeability hemofiltration in septic patients with acute renal failure. Intensive Care Med 2003;29:1989–1995.
17 Bordoni V, Bolgan I, Brendolan A, et al: Caspase-3 and -8 activation and cytokine removal with a novel cellulose triacetate super-permeable membrane in an in vitro sepsis model. Int J Artif Organs 2003;26: 897–905.
18 Morgera S, Haase M, Rocktaschel J, et al: High permeability haemofiltration improves peripheral blood mononuclear cell proliferation in septic patients with acute renal failure. Nephrol Dial Transplant 2003;18:2570–2576.
19 Morgera S, Haase M, Rocktaschel J, et al: Intermittent high-permeability hemofiltration modulates inflammatory response in septic patients with multiorgan failure. Nephron Clin Pract 2003;94:c75–c80.
20 Morgera S, Slowinski T, Melzer C, et al: Renal replacement therapy with high-cutoff hemofilters: impact of convection and diffusion on cytokine clearances and protein status. Am J Kidney Dis 2004;43:444–453.
21 Uchino S, Bellomo R, Morimatsu H, et al: Cytokine dialysis: an ex vivo study. ASAIO J 2002;48:650–653.
22 Lee WCR, Uchino S, Fealy N, et al: β_2-microglobulin clearance with super high flux hemodialysis: an ex vivo study. Int J Artif Organs 2003;26:723–727.
23 Lee WCR, Uchino S, Fealy N, et al: Super high flux hemodialysis at high dialysate flows: an ex vivo assessment. Int J Artif Organs 2004;27:24–28.
24 Lee D, Haase M, Haase-Fielitz A, et al: A pilot, randomized, double-blind, cross-over study of high cut-off versus high-flux dialysis membranes. Blood Purif 2009;28:365–372.
25 Haase M, Bellomo R, Baldwin I, et al: Hemodialysis membrane with a high-molecular-weight cut-off and cytokine levels in sepsis complicated by acute renal failure: a phase 1 randomized trial. Am J Kidney Dis 2007;50:296–304.
26 Parrish WR, Gallowitsh-Puerta M, Czura CJ, et al: Experimental therapeutic strategies for severe sepsis. Ann NY Acad Sci 2008;1144:210–236.
27 Mantell LL, Parrish WR, Ulloa L: HMGB-1 as a therapeutic target for infectious and inflammatory disorders. Shock 2006;25:4–11.
28 Wang H, Bloom O, Zhang M, et al: HMG-1 as a late mediator of endotoxin lethality in mice. Science 1999;285:248–251.
29 Kim JY, Park JS, Strassheim D, et al: HMGB1 contributes to the development of acute lung injury after hemorrhage. Am J Physiol Lung Cell Mol Physiol 2005;288:L958–L965.

Prof. Rinaldo Bellomo
Department of Intensive Care, Austin Hospital
Heidelberg, Victoria, 3084 (Australia)
Tel. +61 3 94965992, Fax +61 3 94963932
E-Mail rinaldo.bellomo@austin.org.au

Author Index

Subject Index